普通高等院校『十四五』规划专门用途英语特色教材

上海市高水平地方应用型大学资金资助建设项目

新能源电力英语课程建设成果

新能源电力英语

主编◎丁肇芬　副主编◎刘法国　徐厚玉

ENGLISH FOR RENEWABLE

Wind, the movement of the air, is the result of temperature differences in different places. Uneven heating results in a difference in atmospheric pressure, which causes the air to move.

ine a system where a heat
y is directly converted to
rical energy. The main
atage of such type of system
higher efficiency is possible
there is no conversion of
y except heat energy to
ical energy.

ELECTRICITY
SOURCES

华中科技大学出版社
http://www.hustp.com
中国·武汉

内 容 提 要

　　本书主要介绍了当今主流新能源电力的基础原理、发展状况、历史及未来发展预测,是新能源电力领域的基础性教材。本书主要面向英语专业本科学生或电力专业各年级需要掌握新能源电力英语的学生,对于其他希望了解新能源电力英语的学者及科研人员也有一定的参考价值。

图书在版编目(CIP)数据

新能源电力英语/丁肇芬主编. —武汉:华中科技大学出版社,2020.12(2023.8重印)
ISBN 978-7-5680-6769-0

Ⅰ.① 新… Ⅱ.① 丁… Ⅲ.① 新能源-电力工业-英语 Ⅳ.① TM7

中国版本图书馆 CIP 数据核字(2020)第 250241 号

新能源电力英语
Xinnengyuan Dianli Yingyu

丁肇芬　主编

策划编辑:宋　焱
责任编辑:王青青
封面设计:廖亚萍
责任校对:封力煊
责任监印:周治超
出版发行:华中科技大学出版社(中国·武汉)　　电话:(027)81321913
　　　　　武汉市东湖新技术开发区华工科技园　　邮编:430223
录　　排:华中科技大学出版社美编室
印　　刷:武汉邮科印务有限公司
开　　本:787mm×1092mm　1/16
印　　张:13　插页:1
字　　数:322千字
版　　次:2023 年 8 月第 1 版第 2 次印刷
定　　价:48.00 元

前言
Preface

随着电力发展步伐的加快和全球面临着能源短缺的困境,新能源在电力开发中越来越引起重视;新能源电力知识对于电力行业的从业人员和翻译人员的重要性日益凸显。对于高等教育和专业教育来说,为了顺应时代潮流,普及新能源电力知识的通识教育也提上了日程。

为了使学生更好地了解新能源电力发展的基础知识和国际前沿信息、理解新能源的历史并展望未来发展,培养更多与国际接轨的跨学科人才,我们编写了此教材。教材介绍了当今主流新能源电力的基础原理、发展状况、历史及未来发展预测,是新能源电力领域的基础性教材。

主要内容:教材第一章分析了新能源的国际消费情况,其后各章分别介绍了太阳能、核能、水能、风能、波浪能、潮汐能、磁流体和生物质能等电力新能源。每章通过思考题和练习题的形式来讨论知识点,特别是发电原理和发展趋势方面的知识,并通过口、笔译实践,帮助学生掌握专业术语和新能源电力的英语表达方法。每章还提供了课外阅读资源,以补充相关的知识或作为扩展阅读材料。

课程总目标是培养学生在电力英语听、说、读、译等方面的基本能力,具体目标包括如下几点。

知识目标:学习新能源电力生产的基本术语,了解其生产过程、历史脉络、未来展望和不同类型新能源发电的优缺点及相关知识。

能力目标:能够读懂有关电力及新能源的英语知识,比较熟练地翻译(口译、笔译)新能源、电网及输电技术的有关材料。

情感目标:爱护地球,保护环境,绿色发展。

　　将来的社会,必将是清洁能源得到重视和大力推广的社会,是全人类健康和谐发展的社会。《新能源电力英语》一书普及新能源基础知识,对于推动可持续发展有重要意义。本书主要面向英语专业本科学生或电力专业各年级需要掌握新能源电力英语的学生,对于其他希望了解新能源电力英语的学者及科研人员也有一定的参考价值。

　　本书参与编写的人员有上海电力大学的丁肇芬、山东沂南马牧池中学的刘法国、浙江大学海洋能实验室的刘浩、山东临沂沂水一中的徐厚玉、山东省淄博第五中学的徐晓东,上海电力大学的余樟亚。本书在编写过程中使用的部分图文,由于客观原因,我们无法与相应作者取得联系,在此一并向相关作者表示感谢。

　　鉴于编者的水平有限,书中难免有疏漏之处,恳请有关专家和读者提出宝贵意见和建议。

编　者
2020 年 4 月

目录
Contents

Unit 1
World Energy Consumption Analysis

Lead-in: *The world is running short of energy. People are aware of the shortage and pay attention to the application of renewable and sustainable energy sources. The analysis in Text A can give us some information on the world energy consumption. Based on the previous energy situation and current environment protection issues, human beings can work together to handle the global warming problem. Text B presents the environmental strategy of China.*

Text A Group Chief Economist's Analysis of World Energy

1. Energy in 2018: an Unsustainable Path

The *Statistical Review of World Energy* has been providing timely and objective energy data for the past 68 years. In addition to the raw data, the *Statistical Review* also provides a record of key energy developments and events through time.

My guess is that when our successors look back at *Statistical Review* from around this period, they will observe a world in which there were growing societal awareness and demands for urgent action on climate change, but where the actual energy data continued to move stubbornly in the wrong direction.

1

新能源电力英语
English for Renewable Electricity Sources

It turns out to be a growing mismatch between hopes and reality. In that context, I fear — or perhaps hope — that 2018 will represent the year in which this mismatch peaked.

Key features of 2018

The headline numbers are the rapid growth in energy demand and carbon emissions. Global primary energy grew by 2.9% in 2018 — the fastest growth seen since 2010 (see Fig. 1-1). This occurred despite a backdrop of modest GDP growth and strengthening energy prices.

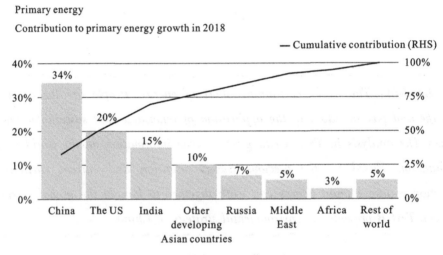

Contributions shown do not sum to 100% due to rounding

Fig. 1-1 Primary world energy growth in 2018

At the same time, carbon emissions from energy use grew by 2.0%, again the fastest expansion for many years, with emissions increasing by around 0.6 **gigatonnes** (see Fig. 1-2). That's roughly equivalent to the carbon emissions associated with increasing the number of passenger cars on the planet by a third.

What drove these increases in 2018? And how worried should we be? Starting first with energy consumption (see Fig. 1-3). As I said, energy demand grew by 2.9% last year. This growth was largely driven by China, the US and India which together accounted for around two thirds of the growth. Relative to recent historical averages, the most striking growth was in the US, where energy consumption increased by a **whopping** 3.5%, the fastest growth seen for 30 years and in sharp contrast to the trend decline seen over the previous 10 years.

Energy demand and carbon emissions

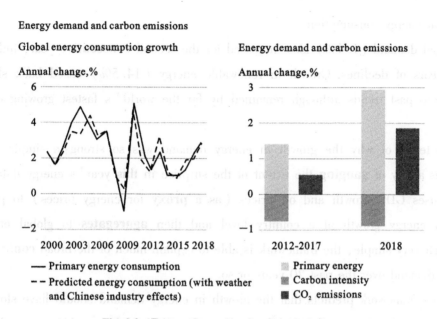

Fig. 1-2   Energy demand and carbon emissions

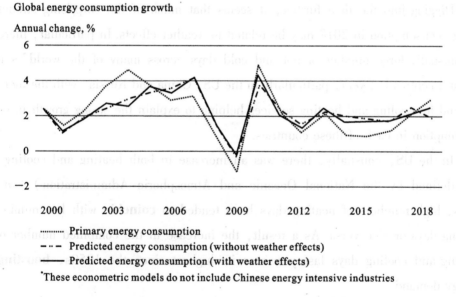

Fig. 1-3   Global energy consumption growth (2000-2018)

The strength in energy consumption was pretty much reflected across all the fuels, most of which grew more strongly than their historical averages. This acceleration was particularly pronounced in natural gas demand, which increased 5.3%, one of its strongest growth rates for over 30 years, accounting for almost 45% of the entire growth in global energy consumption.

Coal demand (1.4%) also increased for the second **consecutive** year, following three years of declines. Growth in renewable energy (14.5%) eased back slightly relative to past trends although remained by far the world's fastest growing energy source.

In terms of why the growth in energy demand was so strong: a simple model provides a way of **gauging** the extent of the surprise in this year's energy data. The model uses GDP growth and oil prices (as a **proxy** for energy prices) to predict primary energy growth at a country level and then **aggregates** to global energy. Although very simple, the framework is able to explain much of the broad **contours** in energy demand over the past 20 years or so.

This framework predicts that the growth in energy demand should have slowed a little last year, reflecting the slightly weaker economic backdrop and the strengthening in energy prices. Instead, energy demand picked up quite markedly.

Digging into the data further, it seems that much of the surprising strength in energy consumption in 2018 may be related to weather effects. In particular, there was an unusually large number of hot and cold days across many of the world's major demand centres last year, particularly in the US, China and Russia, with the increased demand for cooling and heating services helping to explain the strong growth in energy consumption in each of these countries.

In the US, unusually, there was an increase in both heating and cooling days (as defined by the National Oceanic and Atmospheric Administration); in past years, high numbers of heating days have tended to **coincide** with low numbers of cooling days or vice versa. As a result, the increase in the combined number of US heating and cooling days last year was its highest since the 1950s, **boosting** US energy demand.

If we **augment** our framework to include a measure of heating and cooling days for those countries for which data are available, this greatly reduces the extent of the surprise in last year's energy growth. Once these weather effects are included, the growth in energy demand in 2018 still looks a little stronger than expected, but more striking is the surprising weakness of demand growth in the period 2014-2016, which is far lower than the framework predicts.

As discussed in previous *Statistical Review*, much of this appears to stem from some of China's most energy-intensive sectors — iron, steel and cement — which account for around a quarter of China's energy consumption and greatly **dampened** overall energy growth. At the time, I **speculated** that some of the slowing in these sectors reflected the structural rebalancing of the Chinese economy towards more consumer and service-facing sectors and so was likely to persist. But I also noted that the scale of the slowdown suggested that some of it was likely to be **cyclical** and would **reverse** over time. And indeed, that was what began to happen in iron and steel in 2017 and gathered pace in 2018.

If we adjust the framework to also take account of movements in these key Chinese industrial sectors, the over-prediction of energy growth in 2014-2016 is greatly reduced, as is the remaining "unexplained" strength of energy demand in 2018. So, in answer to the question of why energy demand was so strong in 2018: it appears that the strength was largely due to weather related effects — especially in the US, China and Russia — together with a further unwinding of cyclical factors in China.

How does this growth in energy demand relate to the worrying acceleration in carbon emissions?

To a very large extent, the growth in carbon emissions is simply a direct consequence of the increase in energy growth. Relative to the average of the previous five years, growth in energy demand was 1.5 percentage points higher in 2018 and the growth in carbon emissions was 1.4 percentage points higher. One led to the other as the improvement in the carbon intensity of the fuel mix was similar to its recent average.

Finally, in terms of the headline data, what signal might the 2018 data contain for the future?

I think this depends in large part on how you interpret the increasing number of heating and cooling days in 2018. On one hand, if this was just **random variation**, we might expect weather effects in the future to revert to more normal levels, allowing the growth in energy demand and carbon emissions to fall back. On the other hand, if there is a link between the growing levels of carbon in the atmosphere and the types of weather patterns observed in 2018, this would raise the possibility of a worrying **vicious cycle**: increasing levels of carbon leading to more extreme weather patterns, which in

5

turn **trigger** stronger growth in energy (and carbon emissions) as households and businesses seek to offset their effects.

There are many people better qualified than I to make judgements on this. But even if these weather effects are short lived, such that the growth in energy demand and carbon emissions slow over the next few years, the recent trends still feel very distant from the types of transition paths consistent with meeting the Paris climate goals.

So, in that sense, there are grounds for us to be worried.

2. Oil

2018 was another **rollercoaster** year for oil markets (see Fig. 1-4), with prices starting the year on a steady upward trend, reaching the dizzying heights of $85/bbl in October, before **plunging** in the final quarter to end the year at close to $50/bbl.

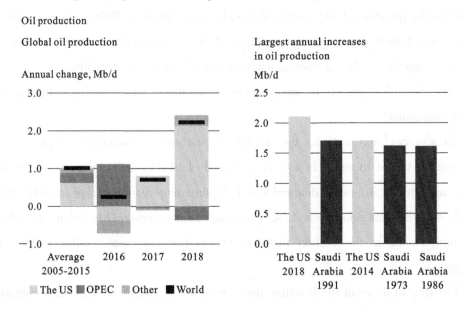

Fig. 1-4 Growth of US oil production, the largest ever annual increase by a single country

Oil demand provided a relatively stable backdrop, continuing to grow **robustly**, increasing 1.4 Mb/d last year. In an absolute sense, the growth in demand was dominated by the developing world, with China (0.7 Mb/d) and India (0.3 Mb/d) accounting for almost two thirds of the global increase. But relative to the past 10 years or so, the big outlier was the US, where oil demand grew by 0.5 Mb/d in 2018, its largest increase for well over 10 years, boosted by increased demand for ethane as new production capacity came on stream. The increased importance of **petrochemicals** in driving oil demand growth was also evident in the global product breakdown, with products most closely related to petrochemicals (ethane, LPG and naphtha) accounting for around half of the overall growth in demand in 2018.

Against this backdrop of steady demand growth, all the excitement came from the supply side, where global production grew by a whopping 2. 2 Mb/d, more than double its historical average. The vast majority of this growth was driven by US production, which grew by 2. 2 Mb/d — the largest ever annual increase by a single country. Since 2012 and the onset of the tight oil revolution, US production (including NGLs) has increased by over 7 Mb/d — broadly equivalent to Saudi Arabia's crude oil exports — an astonishing increase which has transformed both the structure of the US economy and global oil market dynamics. Largely as a consequence, US net oil imports shrunk to less than 3 Mb/d last year, compared with over 12 Mb/d in 2005.

OPEC production fell by 0. 3 Mb/d in 2018, with a marked increase in Saudi Arabian production (0. 4 Mb/d) offset by falls in Venezuela (−0. 6 Mb/d) and Iran (−0. 3 Mb/d). But this year-on-year comparison doesn't do justice to the intra-year twists and turns in OPEC production. The ride began in the first half of 2018 with the continuation of the OPEC + agreement from December 2016. The OPEC + group consistently **overshot** their agreed production cuts during 2017 and this overshooting increased further during the first half of 2018, largely reflecting continuing falls in Venezuelan output. These production cuts helped push **OECD inventories** below their five year moving average for the first time since the collapse in oil prices in 2014.

The first major twist came in the middle of 2018: in response to falling Venezuelan production and the US announcing in May its intention to **impose sanctions on** all Iranian oil exports, the OPEC + group in June committed to achieving 100% **compliance** of their production cuts for the group as a whole.

This commitment contained two important signals. First, given the extent to which production was below the target level, it signalled the prospect of an immediate increase in production. Second, it helped reduce the uncertainty associated with the possibility of future disruptions to either Iranian and Venezuelan production since the commitment to maintain "100% compliance" in essence signalled the willingness of other members of the OPEC+ group to **offset** any lost production.

As a result, between May and November of 2018, net production by the OPEC+ group increased by 900 Kb/d, despite Iranian and Venezuelan production falling by a further 1 Mb/d.

The problem with trying to stabilize oil markets is that there is always some other **pesky** development that you haven't expected. Oil production by Libya and Nigeria — neither of which were part of the OPEC + agreement — increased by more than 500 Kb/d between June and November of last year. As a result, OECD inventories started to grow again. The growing sense of excess supply was compounded by the US announcing in November that it would grant temporary **waivers** for some imports of Iranian oil.

This triggered another twist: a new OPEC + group was formed in December of 2018 — this time excluding Iran and Venezuela, as well as Libya, but including Nigeria — with a **commitment** to reduce production by 1.2 Mb/d relative to October 2018 levels. After a slow start, by the spring of 2019, inventories have fallen back to around their five year average once again.

It's tempting to interpret these twists and turns as **indicative** of OPEC's **waning** powers. But I'm not sure that's the correct interpretation. The role that OPEC+ played in more than offsetting the falls in Iranian and Venezuelan output in 2018 was very significant. For me, the twists and turns simply reflect the difficulty of market management, especially in a world of record supply growth in one part of the world and heightened geopolitical tensions in others. It feels like the rollercoaster will run for some time to come.

3. Natural Gas

2018 was a **bonanza** year for natural gas, with both global consumption and production increasing by over 5% (see Fig. 1-5), one of the strongest growth rates in either gas demand or output for over 30 years. The main actor here was the US, accounting for almost 40% of global demand growth and over 45% of the increase in production.

US gas production increased by 86 bcm, an increase of almost 12%, driven by **shale** gas plays in Marcellus, Haynesville and Permian. Indeed, the US achieved a unique double first in 2018, recording the single largest-ever annual increases by any

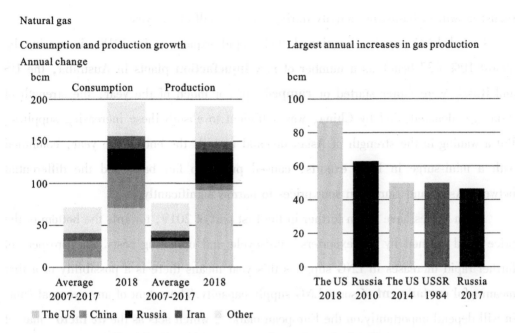

Natural gas

Fig. 1-5 Natural gas consumption and production increase

country in both oil and gas production — in case there was any doubt: the US shale revolution was alive and kicking. The gains in global gas production were supported by Russia (34 bcm), Iran (19 bcm) and Australia (17 bcm).

Although some of the increase in US gas supplies was used to feed the three new US LNG trains which came on stream in 2018, the majority was used to quench the thirst of domestic demand. US gas consumption increased by 78 bcm in 2018 — roughly the same growth as the country achieved over the previous six years. This exceptional strength appeared to be largely driven by the same weather-related effects, with rising demand for space heating and cooling fuelling increased gas consumption, both directly, and, more importantly, indirectly via growing power demand. The expansion of gas consumption within the US power sector was further boosted by almost 15 gigawatts of coal-fired generation capacity being retired in 2018.

Outside of the US, the growth in global gas demand was relatively concentrated across three other countries: China (43 bcm), Russia (23 bcm) and Iran (16 bcm), which together with the US, accounted for 80% of global growth.

China gas consumption grew by an astonishing 18% in 2018. This strength stemmed largely from a continuation of environmental policies encouraging coal-to-gas switching in industry and buildings in order to improve local air quality, together with robust growth in industrial activity during the first half of the year.

Global LNG supplies continued their rapid expansion in 2018, increasing by almost 10% (37 bcm) as a number of new **liquefaction** plants in Australia, the US and Russia were either started or **ramped** up. For much of the year, the strength of Asian gas demand, led by China, was sufficient to absorb these increasing supplies. But a waning in the strength of Asian demand towards the end of the year, combined with a mini-surge in LNG exports, caused prices to fall back and the differential between Asian and European spot prices to narrow significantly.

Asian prices have fallen further in the first part of 2019, towards the bottom of the price band defined by US exporters' full-cycle and operating costs. The prospect of further rapid increases in LNG supplies this year means there is a possibility of a first meaningful **curtailment** of some LNG supply capacity. The extent of any eventual shut-in will depend importantly on the European market, which acts as the **de facto** "market of last resort" for LNG supplies.

Europe's gas demand contracted by a little over 2% (11 bcm) in 2018, but this fall in demand was more than matched (−13 bcm) by continuing declines in Europe's ageing gas fields. The small increase in European gas imports was largely met by LNG cargoes **diverted** from Asia towards the end of the year as the Asian **premium** over European prices almost disappeared.

Russian pipeline exports to Europe were largely unchanged on the year, maintaining the record levels built up in recent years, although with a slight decline in their share of Europe's gas imports. A key factor determining the role that Europe will play in balancing the global LNG market over coming years will be the extent to which Russia seeks to maintain its market share.

4. Coal

China's contribution to global renewables growth, more than the entire OECD combined.

2018 saw a further bounce back in coal — building on the slight pickup seen in the previous year — with both consumption (1. 4%) and production (4. 3%) increasing at their fastest rates for five years. This strength was concentrated in Asia, with India and China together accounting for the vast majority of the gains in both consumption and production.

The growth in coal demand was the second consecutive year of increases, following three years of falling consumption. As a result, the peak in global coal consumption which many had thought had occurred in 2013 now looks less certain: another couple of years of increases close to that seen in 2018 would take global consumption comfortably above 2013 levels.

The growth in coal consumption was more than accounted for by increasing use in the power sector. This is despite continuing strong growth in renewables: renewable energy increased by over 25% in both India and China in 2018, which together accounted for around half of the global growth in renewable energy. But even this was not sufficient to keep pace with the strong gains in power demand, with coal being sucked into the power sector as the balancing fuel.

This highlights an obvious but important point: even if renewables are growing at truly exceptional rates, the pace of growth of power demand, particularly in developing Asia, limits the pace at which the power sector can **decarbonize**.

5. Power Sector and Renewable Energy

The power sector needs to play a central role in any transition to a low carbon energy system: it is the single largest source of carbon emissions within the energy system; and it is where much of the lowest-hanging fruit lies for reducing carbon emissions over the next 20 years (see Fig. 1-6). So, what happened in 2018?

Global power demand grew by 3.7%, which is one of the strongest growth rates seen for 20 years, absorbing around half of the growth in primary energy. The developing world continued to drive the vast majority (81%) of this growth, led by China and India who together accounted for around two thirds of the increase in power demand. But the particularly strong growth of power demand in 2018 owed much to the US, where power demand grew by a **bumper** 3.7%, boosted by those weather effects.

On the supply side, the growth in power generation (see Fig. 1-7) was led by renewable energy, which grew by 14.5%, contributing around a third of the growth; followed by coal (3.0%) and natural gas (3.9%). China continued to lead the way in renewables growth, accounting for 45% of the global growth in renewable power generation, more than the entire OECD combined.

11

Carbon emissions from power sector

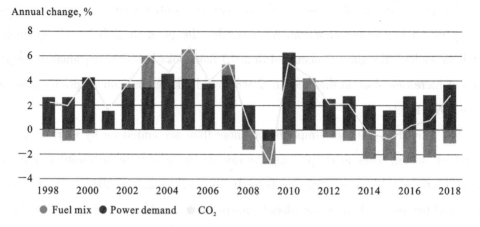

Fig. 1-6　Carbon emissions from power sector

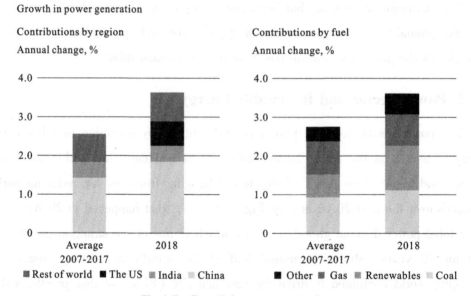

Fig. 1-7　Growth in power generation

Renewable energy appears to be coming of age, but to repeat a point I made in 2018, despite the increasing penetration of renewable power, the fuel mix in the global power system remained depressingly flat, with the shares of both non-fossil fuels (36%) and coal (38%) in 2018 unchanged from their levels 20 years ago.

This persistence in the fuel mix highlights a point that the International Energy Agency (IEA) and others have stressed recently, namely that a shift towards greater electrification helps as a pathway to a lower carbon energy system only if it goes hand-in-hand with a decarbonization of the power sector. Electrification without decarbonizing power is of little use.

In that vein, carbon emissions from the power sector are estimated to have increased by 2.7% in 2018, their highest rate of growth for seven years, accounting for around half of the growth in global carbon emissions. For much of the past 20 years, changes in the carbon intensity of the power sector have been relatively small (or even perverse), such that increases in power demand fed through directly into higher carbon emissions.

6. Electrification without Decarbonization

Over the past five years or so, the rapid growth in renewable energy, together with an edging down in the coal share, has led to a more sustained improvement in the carbon intensity of the power sector, such that the impact of increasing power demand on carbon emissions has been partially offset. But it's still only partial: despite the rapid gains in renewable energy, the pace of growth in power demand has meant that overall carbon emissions from the power sector have increased substantially over the past three years. It hasn't been possible to decarbonize the power sector quickly enough to offset the growth in demand.

To give a sense of the challenge posed by the strength of growth in power demand: if we focus solely on renewable energy, given the profile of demand growth, to maintain the level of carbon emissions from the power sector at its 2015 level, renewable generation would have needed to grow more than twice as quickly than it actually did—by over 1,800 TWh over the past three years, rather than its actual growth of a little over 800 TWh.

A **staggering** number: that additional renewable generation of around 1,000 TWh is roughly equivalent to the entire renewable generation of China and the US combined in 2018.

Alternatively, the same outcome for carbon emissions could have been achieved by replacing around 10% of coal in the power sector with natural gas. The intuition is that renewables are still a relatively small share of power generation relative to coal, and so the proportional movements in coal are a lot smaller.

13

As I mentioned, the general point here is that the robust growth in power demand, particularly in the developing world, greatly adds to the difficulty of decarbonizing the power sector. You have to run very fast just to stand still. This highlights the importance of adopting a range of technologies and fuels, rather than just relying on renewables.

Rapid growth in renewable energy is essential, but it's unlikely to be sufficient. To win the race to Paris, the world is likely to require many fuels and technologies for many years to come.

7. Conclusion

At a time when society is increasing its demands for an accelerated transition to a low carbon energy system, the energy data for 2018 paint a worrying picture, with both energy demand and carbon emissions growing at the fastest rates seen for years.

As I explained, in a statistical sense, it's possible to explain this acceleration in terms of a combination of weather-related effects and an unwinding of cyclical movements in China's pattern of growth. What is less clear is how much comfort we can take from this explanation.

What does seem fairly clear is that the underlying picture is one in which the actual pace of progress is falling well short of the accelerated transition **envisaged** by the Paris climate goals. Last year's developments sound yet another warning alarm that the world is on an unsustainable path.

Source：

https://www.bp.com/content/dam/bp/business-sites/en/global/corporate/pdfs/energy-economics/statistical-review/bp-stats-review-2019-full-report.pdf.

 New words and phrases

....

aggregate	*v.*	总计;合计
	adj.	总数的;总计的
augment	*v.*	增加;提高;扩大
bonanza	*n.*	发财(或成功)的机遇;兴盛;繁荣

boost　*v.*　使……增长;使……兴旺;激励;偷窃;帮助;提高

bumper　*adj.*　丰盛的;丰富的;巨大的

　　　　　n.　缓冲器;减震物;保险杠

　　　　　v.　装满;为……祝酒

coincide　*v.*　同时发生;相符

commitment　*n.*　承诺;许诺;献身

compliance　*n.*　服从;顺从;符合

consecutive　*adj.*　连续不断的

contour　*n.*　外形;轮廓;(地图上表示相同海拔各点的)等高线

curtailment　*n.*　减缩;缩短;限制

cyclical　*adj.*　循环的;周期的

dampen　*v.*　弄湿;使潮湿;抑制,控制,减弱

decarbonize　*v.*　去碳

divert　*v.*　使转向;使绕道;转移

envisage　*v.*　想象;设想;展望

gauge　*v.*　判定,判断(尤指人的感情或态度);(用仪器)测量;估计

gigatonne　*n.*　十亿吨

indicative　*adj.*　陈述的;指示的;象征的

inventory　*n.*　(财产)清单;(商店的)存货,库存

　　　　　v.　开列清单

liquefaction　*n.*　液化;熔解

offset　*v.*　抵消;弥补;补偿

　　　　adj.　胶印的

　　　　n.　开端;出发;平版印刷;抵消;补偿

overshoot　*v.*　超过,越过(预定地点);突破(预计费用)

pesky　*adj.*　麻烦的;讨厌的

petrochemical　*n.*　石油化学产品

plunge　*v.*　暴跌;突降;陡峭地向下倾斜;陷入

　　　　n.　突然跌落;突然分离;(价格、数量的)暴跌;卷入;参与

premium　*n.*　保险费;额外费用;奖金

　　　　　adj.　高昂的;优质的

proxy　*n.*　代理权;代理人;委托书;(测算用的)代替物,指标

ramp	*n.*	斜坡;坡道;活动梯,活动坡道
	v.	敲诈;使有斜面;蔓延
reverse	*v.*	(使)颠倒;撤销,废除(决定、法律等);使反转
	n.	相反的情况或事物;背面;失败
	adj.	相反的;反面的
robustly	*adv.*	粗鲁地;稳健地
rollercoaster	*n.*	过山车;云霄飞车;(情况)忽好忽坏
shale	*n.*	页岩
speculate	*v.*	推测;猜测;投机
stagger	*v.*	摇摇晃晃地走;蹒跚;使震惊;使交错
	n.	摇晃;不稳定
trigger	*v.*	发动;引起;触发;开动;启动
	n.	起因;触发器
waiver	*n.*	(对合法权利或要求的)弃权;弃权声明
wane	*v.*	衰落;衰败;减弱;退潮
whopping	*adj.*	巨大的;惊人的
de facto		(法语)实际上,事实上

the Shah Deniz Alpha platform in the Caspian Sea　里海的阿塞拜疆(气田)平台

OECD：Organization for Economic Cooperation and Development　经济合作与发展组织

OPEC：Organization of the Petroleum Exporting Countries　石油输出国组织

random variation　随机变化;偶然变动

vicious cycle　恶性循环

impose sanction on　强行制裁

 Exercises

Ⅰ. **Read the text and discuss over the following questions with your partner.**

1. How is the situation of world energy consumption?

2. In what orientation should the world shift to energy consumption and production in order to maximize the economic effect and protect the environment? Why?

3. Which countries consume the most fossil fuels in the world, the developed or the developing? How do you know?

4. Which country takes the lead in sustainable energy consumption in the world according to the report? What is the proportion of the country's sustainable energy consumption compared with other countries in the world?

5. What would be the possible result of the accelerating consumption of oil and gas and emission of carbon to the air?

Ⅱ. Fill in the blanks with the words and phrases in the text.

The first major twist came in the middle of 2018: in 1 _____ to falling Venezuelan production and the US announcing in May its intention to 2 _____ sanctions on all Iranian oil exports, the OPEC+ group in June 3 _____ to achieving 100% compliance of their production cuts for the group as a whole.

This 4 _____ contained two important signals. First, given the extent to which production was below the target 5 _____, it signalled the prospect of an immediate 6 _____ in production. Second, it helped reduce the uncertainty 7 _____ with the possibility of future 8 _____ to either Iranian and Venezuelan production since the commitment to maintain "100% compliance" in essence signalled the 9 _____ of other members of the OPEC+ group to 10 _____ any lost production.

Ⅲ. Please give the Chinese or English equivalents of the following terms.

1. Organization for Economic Cooperation and Development (OECD)

2. Organization of the Petroleum Exporting Countries (OPEC)

3. trigger stronger growth

4. coincide with

5. cooling and heating services

6. 碳浓度;碳强度

7. 反之亦然

8. 随机变化;偶然变动

9. 恶性循环

10. 能源需求

IV. Please translate the following sentences into Chinese.

1. This would raise the possibility of a worrying vicious cycle: increasing levels of carbon leading to more extreme weather patterns, which in turn trigger stronger growth in energy (and carbon emissions) as households and businesses seek to offset their effects.

2. The increased importance of petrochemicals in driving oil demand growth was also evident in the global product breakdown, with products most closely related to petrochemicals accounting for around half of the overall growth in demand last year.

3. The problem with trying to stabilize oil markets is that there is always some other pesky development that you haven't expected.

4. Indeed, the US achieved a unique double first last year, recording the single largest-ever annual increases by any country in both oil and gas production.

5. China continued to lead the way in renewables growth, accounting for 45% of the global growth in renewable power generation, more than the entire OECD combined.

6. Rapid growth in renewable energy is essential, but it's unlikely to be sufficient.

V. Please translate the following passage into Chinese.

At a time when society is increasing its demands for an accelerated transition to a low carbon energy system, the energy data for 2018 paint a worrying picture, with both energy demand and carbon emissions growing at the fastest rates seen for years.

As I explained, in a statistical sense, it's possible to explain this acceleration in terms of a combination of weather-related effects and an unwinding of cyclical movements in China's pattern of growth. What is less clear is how much comfort we can take from this explanation.

What does seem fairly clear is that the underlying picture is one in which the actual pace of progress is falling well short of the accelerated transition envisaged by the Paris climate goals. Last year's developments sound yet another warning alarm that the world is on an unsustainable path.

Text B   Environmental Protection in China

1. Foreword

China is a developing country. Now it **is confronted with** the dual task of developing the economy and protecting the environment. Proceeding from its national conditions, China has, in the process of promoting its overall modernization program, made environmental protection one of its basic national policies, regarded the realization of sustainable development as an important strategy and carried out throughout the country large-scale measures for pollution prevention and control as well as **ecological** environment protection.

As a member of the international community, China, while making great efforts to protect its own environment, has taken an active part in international environmental affairs, striven to promote international cooperation in the field of environmental protection, and earnestly **fulfilled** its international **obligations**. All these have given full expression to the sincerity and determination of the Chinese government and people to protect the global environment.

2. The Choice of Implementing a Sustainable Development Strategy

China's modernization drive has been **launched** in the following conditions: the country has a large population base, its **per-capita** average of natural resources is low, and its economic development as well as scientific and technological level remains quite backward. Along with the growth of China's population, the development of the economy and the continuous improvement of the people's consumption level since the 1970s, the pressure on resources, which were already in rather short supply, and on the **fragile** environment has become greater and greater. Which road of development to choose has turned out, historically, to be an issue of **paramount** importance to the survival of the Chinese people as well as their posterity.

The Chinese government has paid great attention to the environmental issues arising from the country's population growth and economic development, and has made protecting the environment an important aspect of the improvement of the people's living standards and quality of life. In order to promote **coordinated** development between the economy, the society and the environment, China enacted and implemented a series of principles, policies, laws and measures for environmental protection in the 1980s.

Making environmental protection one of China's basic national policies. The prevention and control of environmental pollution and ecological destruction and the rational exploitation and utilization of natural resources are of vital importance to the country's overall interests and long-term development. The Chinese government is **unswervingly** carrying out the basic national policy of environmental protection.

Formulating the guiding principles of simultaneous planning, simultaneous implementation and simultaneous development for economic construction, urban and rural construction and environmental construction, and combining the economic returns with social effects and environmental benefits; and carrying out the three major policies of "prevention first and combining prevention with control", "making the causer of pollution responsible for treating it" and " **intensifying** environmental management".

Promulgating and putting into effect laws and regulations regarding environmental protection and placing environmental protection on a legal footing, continuously improving the statutes concerning the environment, formulating strict law-enforcement procedures and increasing the intensity of law enforcement so as to ensure the effective implementation of the environmental laws and regulations.

Persisting in **incorporating** environmental protection **into** the plans for national economic and social development, introducing to it macro regulation and management under state guidance, and gradually increasing environmental protection input so as to give simultaneous consideration to environmental protection and other undertakings and ensure their coordinated development.

Establishing and improving environmental protection organizations under governments at all levels, forming a rather complete environmental control system, and bringing into full play the governments' role in environmental supervision and administration.

Accelerating progress in environmental science and technology. Strengthening research into basic theories, organizing the tackling of key scientific and technological problems, developing and popularizing practical technology for environmental pollution prevention and control, fostering the growth of environmental protection industries, and giving initial shape to an environmental protection scientific research system.

Carrying out environmental publicity and education to enhance the whole nation's awareness of the environment. Widely conducting environmental publicity work, gradually popularizing environmental education in secondary and primary schools, developing on-the-job education in environmental protection and vocational education, and training specialized personnel in environmental science and technology as well as environmental administration.

Promoting international cooperation in the field of environmental protection. Actively expanding exchanges and cooperation concerning the environment and development with other countries and international organizations, earnestly implementing international environmental conventions, and seeking scope for China's role in global environmental affairs.

Since the beginning of the 1990s, the international community and various countries have made an important step forward in exploring solutions to problems of the environment and development. The United Nations Conference on Environment and Development, held in June 1992, made sustainable development the strategy for common development in the future, and this won wide acclaim from the governments of all countries represented at the conference.

In August 1992, shortly after that conference, the Chinese government put forward ten major measures China was to adopt to enhance its environment and development, clearly pointing out that the road of sustainable development was a logical choice for China now and in the future.

In March 1994, the Chinese government approved and promulgated *China's Agenda 21 — White Paper on China's Population, Environment, and Development in the 21st Century*. This document, proceeding from the country's specific national conditions in these three respects, put forward China's overall strategy, measures and program of action for sustainable development. The various departments and localities also worked out their respective plans of action to implement the strategy for sustainable development.

At its Fourth Session in March 1996, China's Eighth National People's Congress examined and adopted the Ninth Five-Year Plan of the People's Republic of China for

National Economic and Social Development and the Outline of the Long-Term Target for the Year 2010. Both the Plan and Outline take sustainable development as an important strategy for modernization, thus making it possible for the implementation of the strategy of sustainable development in the course of China's economic construction and social development.

China pays great attention to environmental **legislative** work and has now established an environmental **statutory** framework that takes the Constitution of the People's Republic of China as the foundation and the Environmental Protection Law of the People's Republic of China as the main body.

China has enacted and promulgated many special laws on environmental protection as well as laws on natural resources related to environmental protection. They include the Law on the Prevention and Control of Water Pollution, Law on the Prevention and Control of Air Pollution, Law on the Prevention and Control of Environmental Pollution by Solid Wastes, Marine Environment Protection Law, Forestry Law, Grassland Law, Fisheries Law, Mineral Resources Law, Land Administration Law, Water Resources Law, Law on the Protection of Wild Animals, Law on Water and Soil Conservation, and Agriculture Law.

The National People's Congress has established an Environment and Resources Protection Committee, whose work is to organize the formulation and examination of drafted laws related to environmental and resources protection and prepare the necessary reports, exercise supervision over the enforcement of laws governing environmental and resources protection, put forward motions related to the issue of environmental and resources protection, and conduct exchanges with parliaments in other countries in the field of environmental and resources protection. The people's congresses of some provinces and cities have also established corresponding environmental and resources protection organizations.

The Chinese government regards prevention and control of industrial pollution as the focal point of environmental protection. Thanks to **unremitting** efforts over the past 20-odd years, China has made great progress in this regard.

Changes in the strategy for the prevention and control of industrial pollution have been effected. In the 1970s, efforts to prevent and control industrial pollution in China mainly concentrated on the control of point sources. In the 1980s, China carried out prevention and control of industrial pollution in a comprehensive way through the readjustment of irrational industrial distribution, the overall industrial structure and the product mix in combination with technical transformation, strengthened environmental management and other policies and measures. In the course of founding the socialist market economic system in the 1990s, China has changed its traditional development strategy, promoted clean production and **embarked on** the sustainable development road. In guiding concept for the prevention and control of industrial pollution, "three changes" have been decided upon, i. e. , regarding basic strategy, China will gradually change its strategy of end-of-pipe pollution control into pollution control during the whole process of industrial production; with respect to the control of pollutant discharge, concentration control will be replaced by a combination of the control of concentration and that of total quantity; and with regard to pollution control methods, focus on the control of scattered point sources will be replaced by a combination of centralized and scattered controls.

Source:

https://www.fmprc.gov.cn/ce/celt/eng/zt/zfbps/t125245.htm.

 New words and phrases

accelerate   *v.*   (使)加速;加快

coordinate   *v.*   使协调;使相配合

ecological   *adj.*   生态的;生态学的

fragile   *adj.*   易碎的;脆弱的;不牢固的;精细的

fulfill   *v.*   履行,执行,贯彻;完成

intensify   *v.*   (使)加强;增强;加剧

launch   *v.*   开始从事,发起;(首次)上市,发行;使(船,尤指新船)下水

　　　　 n.   发射;发行;下水

legislative   *adj.*   立法的;制定法律的

　　　　　 n.   立法权;立法机关

obligation   *n.*   义务;职责;责任

paramount   *adj.*   至关重要的;首要的;至高无上的

per-capita   *adj.*   人均的

promulgate   *v.*   传播;宣传;宣布;颁布,发布(新法律或体制)

statutory   *adj.*   法定的;法令的;法定代理的

unremitting   *adj.*   不停的;持续不断的;不懈的

unswervingly   *adv.*   坚定不移地

be confronted with   面对;面临着

embark on   着手,开始

incorporate. . . into   合并……入内;并入

Exercises

Please think over the following questions and try to answer them according to the text.

1. What is China's attitude toward environment protection?

2. What strategies has Chinese government taken to protect the environment?

3. What are the "three changes" in the prevention and control of industrial pollution?

Keys to Exercises

Unit 2
Solar Energy

Text A   Solar Power on the Earth

Solar energy is the most important energy of various renewable energy, and also the most abundant energy that human beings can use. Solar energy has been applied to various domains: solar cooker, solar car, solar power station, unmanned weather stations, communication stations, television **relay stations**, sun clock, power rails, **black light, beacon lights**, railway signals, etc. Solar energy is so widely used that we can never exaggerate its significance. In recent years, solar energy has played a major role in the global energy transformation trend. Solar energy utilization is expected to be increasingly important in order to meet future energy needs and limit CO_2 emissions to the atmosphere. The text will focus on the primary knowledge of solar

energy as to what it is, how electricity is produced from solar energy, the application, the history and world distribution of solar power, the benefits, the prospect and related issues.

1. What Is Solar Energy?

Solar energy is radiant light and heat from the Sun **harnessed** by using a range of ever-evolving technologies such as solar heating, **photovoltaics**, solar thermal energy, solar architecture and **artificial photosynthesis**.

Scientists have said that the number of **photons** that reach the Earth in just one hour, would be enough to power the world for a whole year, and we would never need to use fossil fuels again, theoretically speaking. Practically, the surface of the Earth varies a lot in receiving solar energy. Variation of solar energy resource is primarily caused by **aerosol** and cloud effects. In 2013, Calinoiu et al. in Romania showed the aerosol pollution events were estimated to cut off collectable solar energy over 20%. Further, the research group concluded that the aerosol pollution episodes should be considered in both development and exploiting stage of solar energy collection.

2. The Work Principle

Solar energy is an important source of renewable energy and its technologies are broadly characterized as either passive solar or active solar depending on the way they capture and distribute solar energy or **convert** it **into** solar power. Active solar techniques include the use of photovoltaic systems, concentrated solar power and solar water heating to harness the energy. Passive solar techniques include orienting a building to the Sun, selecting materials with favorable thermal mass or light dispersing properties, and designing spaces that naturally circulate air.

Solar power systems work by converting the Sun's **electromagnetic** energy into either solar thermal energy or solar electricity. Solar electric systems create solar electricity using solar power panels. The DC electricity generated by the solar panels is converted to AC current and can be used by all of our household appliances.

2.1　How do solar panels work?

It is important to begin by mentioning that a **solar panel** is different in scale and applications from a solar cell. In fact, such a panel is an assembly of multiple photovoltaic cells. While smaller solar cells are used to power everyday objects like calculators, the bigger and rarer solar panels are used in almost **exotic** ways — in emergency road signs, buoys, **parking lots** — in order to provide power to the lights. The process by which solar cells convert the energy from the Sun directly into power is interesting and helps explain why more research needs to be carried out before the process can become cost-effective.

2.2   Process of conversion of photons into electrons

We will now explain the process through which light gets converted to electricity in a solar panel, by taking the solar cell as a unit of observation. Once the Sun's light hits certain materials, it forms an electrical current which is how solar energy is created. Solar cells, also popularly known as photovoltaic cells, help convert the energy **derived from** the Sun directly into charge. Photovoltaic cells are composed of a unique material known as silicon (see Fig. 2-1). Basically, what happens is when the light hits the cell, a specific part of it gets absorbed by the material of the **semiconductor**. **Electrons** get knocked loose due to the energy and this gives them the ability to **drift** freely.

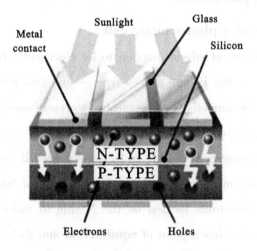

Fig. 2-1   Solar panels and silicon

https://www.solarenergybase.com/how-do-solar-panels-work/#more-130.

All photovoltaic cells posses one or several electric fields that force the loosely flowing electrons to drift in a particular direction. This **electron flow** happens to be a current which can be drawn off for use externally, by inserting metal contacts both on the bottom and top of the cell. Together with the voltage of the cell, this current defines the amount of power that the photovoltaic cell is capable of emitting.

2.3　Use of silicon in solar cells

Silicon possesses a few unique chemical traits, especially when it is still in the form of a crystal.

· A silicon atom contains 14 electrons which are set in three separate shells.

· The initial two remain completely filled.

· Only four electrons fill half of the outer shell.

· An atom of silicon is always trying to complete the last shell and it shares electrons from four **adjacent** atoms to achieve this.

· This leads to the development of a crystalline structure which is vital for such a solar cell.

However, pure silicon in crystalline form is a bad electricity conductor since the electrons are unable to move freely. Therefore, the silicon present in a photovoltaic cell contains **impurities** that allow the cell to work. **Phosphorus**, forming N-type (negative) silicon, and **boron**, forming P-type (positive) silicon, are commonly added as impurities to pure silicon.

2.4　Structure of a photovoltaic cell

Two individual pieces of silicon happen to be electrically **neutral**. But when two impure silicon pieces — the P-type and N-type — are brought in contact with each other, an electric field is formed. The free electrons present on the N side seeing all the holes on the P side rush to fill them fast. All free electrons are unable to fill the free openings. The entire arrangement would, in fact, be wasted if they had. But, exactly at the opening point, they mix successfully to form a kind of **barrier** which makes it increasingly **tough** for electrons located on the N side to make the move to side P. Gradually, the process reaches a point of **equilibrium** and there exists an electric field dividing the different sides. This field serves as a kind of **diode**, permitting and even forcing electrons to drift from the P to the N side, instead of the other way round.

When light in proton form hits the solar cell, the energy results in dividing the hole-electron couples. Each photon carrying sufficient amount of energy will usually be able to free just a single electron, leading to a free opening as well. If this occurs very close to the field of electricity, or if free opening and free electrons tend to flow into influence range, the field is likely to shift the electron to side N and the hole to side P.

This results in even more **disruption** of neutrality of electricity, and if an external charged path is presented, electrons make use of the path to travel to the P side to combine with openings sent by the electric field, carrying on work in the process. The flow of electron results in the current and the electric field of the cell results in the formation of a voltage. Power is formed as a product of both voltage and current (see Fig. 2-2).

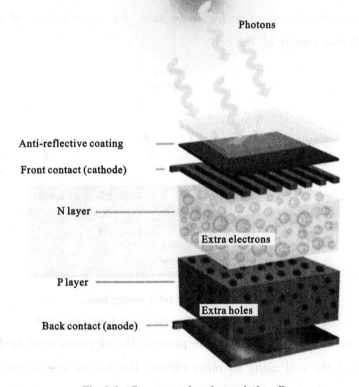

Photons

Anti-reflective coating

Front contact (cathode)

N layer

Extra electrons

P layer

Extra holes

Back contact (anode)

Fig. 2-2   Structure of a photovoltaic cell

A few elements remain before the cell can be used. Silicon is a highly shiny substance that can push photons away **prior to** the completion of their job. In order to **remedy** this, it is normal to apply an **anti-reflective coating** which lessens the losses. The ultimate step is the installation of a cover plate made of glass or any other object that can offer protection to the cell from the elements. The amount of energy absorbed from sunlight by the solar cell is not very much.

2.5 Loss of energy in solar cells

It is possible to divide light into many wavelengths. The light that affects the cell contains photons consisting of a large variety of energies. Not all of them have energy enough to modify the opening-electron pair. However, there are other electrons that contain lots of energy. Only a specific portion of this energy, measured in terms of eV or **electron volts** is necessary to knock an extra electron loose. This is commonly known as a material's band gap energy. If a photon contains excess energy than what is required, the additional energy gets lost (see Fig. 2-3). However, if the **incremental** energy is equal to the required amount, there is possibility of the formation of more than one electron-hole couple. But, the latter effect does not seem to be of much significance. Just these two outcomes account for more than 70 percent loss of the incident concerning radiation energy on the cell.

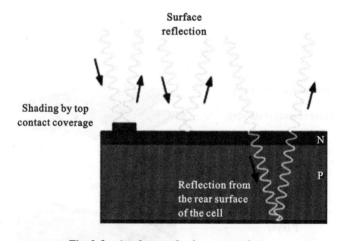

Fig. 2-3 A scheme of solar energy losses

https://www. solarenergybase. com/how-do-solar-panels-work/#more-130.

There are more losses involved in the process. The electrons need to drift from one side to another of the cell using an outer circuit. The bottom can be covered using a metal, leading to effective conduction. But in case the top is completely covered, the photons are unable to pass through the **opaque** conductor and lose most of their current.

In order to reduce these losses, cells are normally covered using a **contact grid** made of metal that lessens the distance required to travel by the electrons, covering just a limited part of the surface of the cell. Even then, the grid blocks a few electrons.

3. Application of Solar Panels

3.1   BIPV

Building-integrated photovoltaics (**BIPV**) are photovoltaic materials that are used to replace conventional building materials in parts of the building envelope such as the roof, skylights, or **facades**. They are increasingly being incorporated into the construction of new buildings as a principal or **ancillary** source of electrical power, although existing buildings may be **retrofitted** with similar technology. The advantage of integrated photovoltaics over more common non-integrated systems is that the initial cost can be offset by reducing the amount spent on building materials and labor that would normally be used to construct the part of the building that the BIPV modules replace. These advantages make BIPV one of the fastest growing segments of the photovoltaic industry.

The PV industry requires realistic statistics of solar resource during the development period (e. g. plant location) and also anticipation for possible variations of this resource during exploitation stage (energy storage). The most common parameter used to express solar resource is **the global horizontal irradiance** (GHI) (Wm^{-2}), which is the total amount of solar energy falling on a horizontal surface per unit area per unit time integrated over the solar spectrum. The GHI is the sum of the direct normal irradiance (DNI) and **the diffuse horizontal irradiance** (DHI).

3.2   Solar air-conditioning system

Solar energy refrigeration and air conditioning usually use solar collectors and adsorption refrigerators combined (see Fig. 2-4). The collector is used to provide heat generator needed for the refrigerator, so as to make the refrigerator achieve larger **coefficient of performance** (COP).

Johnson Space Center's solar-powered refrigeration system employs a PV panel, vapor compressor, thermal storage and reservoir, and electronic controls. The process that makes the refrigeration possible is the conversion of sunlight into **DC electrical power**, achieved by the PV panel. The DC electrical power drives the compressor to circulate **refrigerant** through a vapor compression refrigeration **loop** that extracts heat from an insulated enclosure. This enclosure includes the thermal reservoir and a **phase**

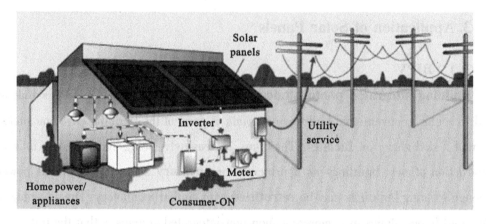

Fig. 2-4　Solar air-condition system

https://1stle.com/solar-energy-basics/.

change material. This material freezes as heat is extracted from the enclosure. This process effectively creates an "ice pack," enabling temperature maintenance inside the enclosure in the absence of sunlight. Proper sizing of the highly **insulated cabinet**, phase change thermal storage, variable speed compressor, and solar PV panel allow the refrigerator to stay cold all year long. To optimize the conversion of solar power into stored thermal energy, a compressor control method fully exploits the available energy. Other power optimization measures include: smoothing the power voltage via a **capacitor**, providing additional current during compressor start-up, monitoring the rate of change of the smoothed power voltage using a controller to determine if the compressor is operating below or above the available power maximum, enabling adjustment of the compressor speed if necessary, replacing the **capillary tube** in the refrigerator system with an expansion **valve**, improving energy efficiency in certain operating conditions. These adjustments to the compressor operation contribute to the conversion of the majority of the available solar power into stored thermal energy. Applications may include a cold side water loop or incorporation of the evaporator into the thermal storage. Electronic controls can also be added to provide **backup power** from an alternative power source such as an electric grid.

3.3　Solar power plant

Components can also form a solar power plant, which can be used for different power applications. Solar photovoltaic systems usually include batteries, storage, **inverter**, and control. The application of the building can be divided into stand-alone

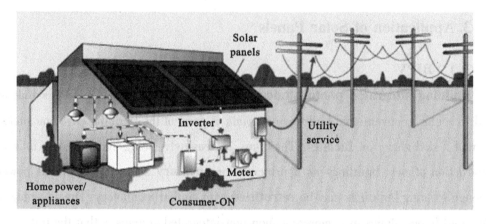

photovoltaic system and grid connected photovoltaic system. The application of solar cells provides lighting and other power for buildings. If solar energy can meet the energy demand of buildings, it can be called "zero energy house".

At present, most of the ways of utilizing solar energy in China are solar photovoltaic conversion, such as solar water heater, solar oven, solar room, solar drying, solar greenhouse, solar refrigeration and air conditioning, solar power generation and photovoltaic power generation system.

4. The History

As early as 1839, at the age of 19, Becquerel, France, conducted physical experiments and found that when light shone on two kinds of metal electrodes in a conductive liquid, the electric current would increase, thus discovering the "photovoltaic effect".

During the period from 1900 to 1920, the focus of solar energy research in the world was still the solar power plant, but the way of concentrating light was diversified and the plate collector and low-boiling working medium were used. The device gradually expanded, with the maximum output power reaching 73. 64 kW. Typical devices are as follows: in 1901, a solar water pumping device was built in California, the United States, with a truncated conical concentrator, with power of 7. 36 kW; from 1902 to 1908, five sets of dual-cycle solar energy engines were built in the United States, with plate collector and low-boiling working medium; in 1913, a solar water pump consisting of five parabolic trough mirrors was built south of Cairo, Egypt, with a total lighting area of 1,250 square meters.

From 1945 to 1965, in the 20 years after the end of the Second World War, solar energy research was gradually resumed and carried out, and solar energy academic organizations were established to hold academic exchanges and exhibitions, which once again gave rise to the **upsurge** of solar energy research. At this stage, some significant advances have been made in solar energy research. Among them, the practical silicon solar cell developed in 1945 laid the foundation for large-scale application of photovoltaic power generation. In 1955, Israel Taber et al. proposed the basic theory of selective coating at the first international solar thermal scientific conference, and developed practical selective coating such as black nickel, which created conditions for the development of efficient collector. In 1952, the French national research center built a 50 kW solar furnace in the eastern **Pyrenees**. In 1960, the world's first set of **flat plate collector** heating **ammonia-water** absorption air conditioning system was built in Florida. A stirling engine with quartz windows was introduced in 1961. At this stage, the research on solar energy basic theory and basic materials was strengthened, and great breakthroughs were made in technologies such as solar selective coating and silicon solar cells. Plate collector achieved great development with the technology gradually mature. Progress was made in the study of solar absorption air conditioners, and a number of experimental solar houses were built.

The oil crisis from 1973 to 1980 made people realize that the existing energy structure must be completely changed and the transition to the future energy structure should be accelerated. As a result, many countries, especially the industrialized countries, had re-strengthened their support for the development of solar energy and other renewable energy technologies, and the development and utilization of solar energy was once again on the rise in the world. During this period, the development and utilization of solar energy was in an unprecedented period of great development. Countries strengthened the planning of solar energy research, and many had formulated short-term and long-term sunshine plans. Development and utilization of solar energy had become a government action, and support had been greatly strengthened. International cooperation became very active. Some third world countries had taken an active part in the development and utilization of solar energy. The research field was expanding; the research work was deepening, and a lot of great achievements had been achieved, such as **vacuum collector tube**, **amorphous** silicon solar cell, **photodissociated** water to produce hydrogen, and etc. However, the solar energy development plans formulated by various countries generally had the problems of too high requirements in urgency, and the difficulties in the implementation process were underestimated.

In 2011, the International Energy Agency said that "the development of affordable, **inexhaustible** and clean solar energy technologies will have huge longer-term benefits. It will increase countries' energy security through reliance on an **indigenous**, inexhaustible and mostly import-independent resource; meanwhile, it will enhance sustainability, reduce pollution, lower the costs of mitigating global warming, and keep fossil fuel prices lower than otherwise. These advantages are global. Hence the additional costs of the **incentives** for early deployment should be considered learning investments; they must be wisely spent and need to be widely shared.

5. World Distribution

The Sun's radiation reaches the Earth in a **non-homogeneous** way because of its interaction with the atmosphere and **the angle of incidence** of sunrays. The angle of incidence varies according to two factors: the Earth's rotation around its axis, which is very important for the alternation of day and night, and the **inclination** of the Earth's axis as compared to **the plane of** its **orbit**, leading to a seasonal variation of the maximum height of the Sun on the horizon. On one hand, when the Sun **is perpendicular to** the Earth's surface, the maximum concentration of sunrays on the ground is obtained. On the other hand, if the sunrays reach the Earth's surface with a certain inclination, the same amount of energy is **dispersed** over a larger surface. Therefore, solar energy can be highly exploited only within a belt included between 45° **latitude** south and north. For convenience and simplicity, the geographic distribution of total solar radiation on a global scale is divided in terms of intensity into four broad belts around the Earth. These are described briefly with respect to the northern **hemisphere**, with the understanding that the same conditions apply to the corresponding belts in the southern hemisphere.

The most favourable belt. This belt, lying between latitudes 15°N, and 35°N, embraces the regions that are naturally **endowed** with the most favourable conditions for solar energy applications. These **semi-arid** regions are characterized by having the greatest amount of solar radiation, more than 90% of which comes as direct radiation because of the limited cloud coverage and rainfall (less than 250 mm per year). Moreover, there is usually over 3,000 hours of sunshine per year.

Moderately favourable belt. This belt lies between the equator and latitude 15°N, and is the next most favourable region for the purpose previously mentioned. Because the humidity is high, and cloud cover is frequent, the proportion of scattered radiation is quite high. There is a total of about 2,500 hours of sunshine per year. The solar intensity is almost **uniform** throughout the year as the seasonal variations are only slight.

Less favourable belt. This belt lies between latitude 35°N and 45°N. Although the average solar intensity is roughly about the same as for the other two belts, there are marked seasonal variations in both radiation intensity and daylight hours. During the winter months, solar radiation is relatively lower than in the rest of the year.

Least favourable belt. The regions in this belt lie beyond latitude 45° N. They include Russia, and the greater parts of northern Europe and North America. Here, about half of the total radiation is diffuse radiation, with a higher proportion in winter than in summer primarily because of the rather frequent and extensive cloud coverage.

Distribution in China

China is one of the countries with abundant solar energy resources. The annual sunshine hours are more than 2,000 hours and the annual radiation amount is more than 5,000 MJ/m^2 in the area over 2/3 of the country. According to the statistical analysis, the total solar radiation received by China's land area every year is 3. 3×103-8. 4×103 MJ/m^2, which is equivalent to 2. 4 × 10. 4 billion tons of standard coal reserves.

China's solar energy resource areas are divided into four categories.

Area 1 (rich resource zone): The annual radiation **dose** is 6700-8370 MJ/m^2, equivalent to 230 kg of standard coal burning heat. It mainly includes **Qinghai-Tibet Plateau**, northern Gansu, northern Ningxia, southern Xinjiang, northwest Hebei, northern Shanxi, and other places.

Area 2 (abundant resource zone): The annual radiation amount is 5400-6700 MJ/m^2, which is equivalent to the heat from 180-230 kg standard coal **combustion**. It mainly includes Shandong, Henan, southeast Hebei, south Shanxi, north Xinjiang, Jilin, Liaoning, Yunnan, north Shanxi, southeast Gansu, south Guangdong, south Fujian, north central Jiangsu and north Anhui.

Area 3 (general resource zone): The annual radiation dose is 4200-5400 MJ/m^2, which is equivalent to 140-180 kg of standard coal burning heat. It is mainly in the middle and lower reaches of the Yangtze river, part of Fujian, Zhejiang and Guangdong.

Area 4 (poor resource zone): The annual radiation exposure is less than 4200 MJ/m^2. It mainly includes Sichuan and Guizhou provinces. Those areas have the least solar energy resources in China.

Tibet has the most abundant solar radiation energy in China. The annual total radiation in the whole region is mostly between 5000-8000 MJ/m^2, and the distribution is increasing from east to west.

6. Advantages and Disadvantages

For climates with moderate heating and cooling needs, heat pumps offer an energy-efficient alternative to furnaces and air conditioners. During the heating season, heat pumps move heat from the cool outdoors into the warm house and during the cooling season, heat pumps move heat from the cool house into the warm outdoors. Because they move heat rather than generate heat, heat pumps can provide equivalent space conditioning at as little as one quarter of the cost of operating conventional heating or cooling appliances. The advantages of solar energy mainly include but are not restricted to the followings.

(1) This kind of energy is almost endless. This is a useful and long-term energy source for human beings in the future. It can be exploited in all regions of the world with available reserves every day. As long as the Sun remains, we can still approach solar energy.

Energy sources such as wind, water, coal, and etc. need fuels to be converted into electricity through the hydropower and thermal power plants. Of course, these materials are limited. Unlike them, solar energy does not require any fuel. It is an infinite and renewable source of energy.

(2) It is environmentally friendly. Solar energy is a green energy source and environmentally friendly. It does not emit any carbon, nitrogen oxide, sulfur gas, mercury, or any gas that is harmful to the environment. Especially it is 100% safe to human work when generating electricity. So solar power generation does not contribute to global warming, acid rain or fog. In fact, it contributes to reducing harmful greenhouse gas emissions. Since solar energy does not use any fuel in the electricity generation process, there are no storage, transportation or retrieval of fuel. It can be said that this is considered the best source of clean energy that nature gives us.

(3) Solar energy is free. It is produced free of charge. Solar energy does not require any fuel, so its price is not affected by supply and demand on the market. That is why using solar energy is a relatively economical solution.

(4) Low maintenance costs. Solar systems often last for decades, so they do not require much maintenance. The reason is that the solar system has no moving parts so there will be no wear and tear. The only need is to keep them clean and replace the converter every 5 years. There is only little maintenance that is needed to keep the solar cells running.

(5) Diverse applications. Solar energy can be very helpful in giving power to calculators and other low power consuming devices. This kind of energy can be used even in remote areas.

(6) Solar energy does not create much noise. Compared with other energy sources, solar panels are silent producers.

The disadvantages of solar energy mainly include the followings.

(1) It may have high initial investment and storage costs. They are very expensive to build and some of the solar power stations do not actually match the power output of similar sized conventional power stations.

(2) It is dependent on some factors such as the seasons of the year, the time of the day and the location. Solar energy requires the energy of the Sun and because of that it cannot be created during rainy days or at night and is limited to some of the countries' climate.

(3) High demand for interior space. The installation requires large space in the home. Solar panels take a lot of space. Not all roofs are large enough to house the required number of solar panels.

7. Future Development

The future of solar energy concerns only the two widely recognized classes of technologies for converting solar energy into electricity — photovoltaics (PV) and concentrated solar power (CSP), sometimes called solar thermal — in their current and **plausible** future forms. Because energy supply facilities typically last several decades, technologies in these classes will dominate solar-powered generation between now and 2050, and we do not attempt to look beyond that date.

We concentrate on the use of grid-connected solar-powered generators to replace conventional sources of electricity. For the more than one billion people in the developing world who lack access to a reliable electric grid, the cost of small-scale

PV generation is often outweighed by the very high value of access to electricity for lighting and charging mobile telephone and radio batteries. In addition, in some developing nations it may be economic to use solar generation to reduce reliance on imported oil, particularly if that oil must be moved by truck to remote generator sites. A companion working paper discusses both valuable roles for solar energy in the developing world.

Research is advancing on materials that are used to harness the energy from the Sun and ease the future life of human kind. The challenge is to design the state of the art solar energy materials with **optimum optical** properties and to operate them in the wide wavelength range of visible **spectrum**. The materials can be varied from organic to inorganic semiconductors of **polymers** and metals / non-metals, which can be in the form of single to **multilayer** thin films or bulk materials.

In spite of the importance of solar statistics, ground-based solar irradiance observation sites are not widely spread around the world. Most solar irradiance data are based on satellite-observations combined with modeling and have large spatial coverage. Such estimates need to be complemented with accurate ground measurements to improve their local accuracy, since satellite **algorithms** do not take into account local effects due to e. g. high mountains, desert, and snow.

Common solar power issues

A portion of the power generated by the photovoltaic panels can be collected using chemical batteries. However, the setup suffers from a lack of additional power initially. The photon-producing sunlight is also responsible for **infrared** and harmful ultraviolet waves which result in the physical degradation of the panels. The panels, need to be exposed to the **harsh** elements of nature, are also capable of affecting efficiency in a serious manner.

The necessity of light in the emission of electrical current can be deduced from the name of the photovoltaic cell. Future scientists will face the challenge of creating regular-sized and further developed solar panels that can be used to produce additional energy for the times when there is no sunlight.

Apart from **aforementioned** issues, there is also the potential hourly over- and under-generation issue in the near future.

Source：

1. https：//www. solaralliance. org/how-does-solar-energy-work/.

2. https：//www. solarenergybase. com/how-do-solar-panels-work/#more-130.

3. https：//1stle. com/solar-energy-basics/.

4. https：//airneeds. com/advantages-and-disadvantages-of-solar-energy/.

5. https：//apecsec. org/advantages-and-disadvantages-of-solar-energy/.

 New words and phrases

aforementioned *adj.* 前面提到的；上述的

adjacent *adj.* 与……毗连的；邻近的

aerosol *n.* 喷雾器，雾化器；气雾剂；气溶胶

algorithm *n.* 算法；计算程序

ammonia-water *n.* 氨水

amorphous *adj.* 无定形的；无组织的；非晶形的

ancillary *adj.* 辅助的；补充的

barrier *n.* 屏障；障碍物；阻力；隔阂

boron *n.* 硼；硼粉

capacitor *n.* 电容器；电容

combustion *n.* 燃烧

diode *n.* 二极管

disperse *v.* （使）分散，（使）散开；散播；使（光）色散

  adj. 分散的

disruption *n.* 分裂，瓦解；破裂，毁坏；中断

dose *n.* 剂量，药量；（药的）一服，剂

  v. 服药

drift *v.* 流动；随意移动；浮现

  n. 漂移；趋势，动向

electron *n.* 电子

electromagnetic *adj.* 电磁的

endow *v.* 捐赠，资助；赋予，赐予

equilibrium *n.* 平衡，均势；平静

exotic   *adj.*   异国的;外来的;异乎寻常的

            n.   舶来品;外来物

facade   *n.*   建筑物的正面;外表;虚伪,假象

harness   *v.*   利用;给(马等)套轭具;控制

harsh   *adj.*   粗糙的;刺耳的;严厉的,严格的;残酷的

hemisphere   *n.*   半球

impurity   *n.*   不纯;不洁;杂质

incentive   *n.*   激励;奖励;诱因;奖励措施

inclination   *n.*   倾向;爱好;斜坡

incremental   *adj.*   增加的;逐步的

indigenous   *adj.*   土生土长的;生来的,固有的

inexhaustible   *adj.*   无穷无尽的;用不完的

infrared   *adj.*   红外线的

            n.   红外线

inverter   *n.*   反用换流器,变极器;反相器;变频器

latitude   *n.*   纬度;范围;界限

loop   *n.*   圈,环;回路;弯曲部分

            v.   (使)成环,(使)成圈;以环连结;使翻筋斗

multilayer   *n.*   多层;多层次

neutral   *adj.*   中立的;(化学中)中性的;暗淡的;不带电的

non-homogeneous   *adj.*   不均匀的;非同质的

opaque   *adj.*   不透明的;晦暗的;不传导性的;含糊的

            n.   不透明,晦暗;遮檐;遮光涂料

phosphorus   *n.*   磷;磷光体

photodissociated   *adj.*   光电离的

photon   *n.*   光子;光量子

photovoltaics   *n.*   太阳光电;太阳能光电板;光伏发电,光伏技术

plausible   *adj.*   貌似真实的;貌似有理的

polymer   *n.*   聚合物(体)

refrigerant   *n.*   制冷剂;冷冻剂

            adj.   冷却的;制冷的

remedy	n.	治疗法;补救办法;纠正办法
	v.	改正,纠正,改进;补救;治疗

retrofit   v.   给机器设备装配(新部件);翻新;改型

semi-arid   adj.   半干旱的

semiconductor   n.   半导体

silicon   n.   硅;硅元素

spectrum   n.   光谱;波谱;范围;系列

tough   adj.   坚强的,坚韧的,不屈不挠的;艰苦的;难办的;牢固的;强壮的;
粗暴的

uniform   adj.   统一的,一致的;均衡的;始终如一的

         n.   制服

upsurge   n.   高涨;高潮

         v.   高涨;涌起

valve   n.   阀;真空管

         v.   装阀于;以活门调节

anti-reflective coating   防反射涂层

artificial photosynthesis   (人工)光合作用

backup power   备用电源

beacon light   灯塔

be perpendicular to   垂直于

building-integrated photovoltaics (BIPV)   光伏建筑一体化

black light   不可见光;黑光(指紫外线和红外线)

capillary tube   毛细管

coefficient of performance (COP)   制冷系数

contact grid   相接电网

convert... into...   把……转化为

DC electrical power   直流电源

derive from   采自;来源于

electron flow   电子流

electron volt   电子伏特

flat plate collector   平板集热器

insulated cabinet   保温箱;隔热箱

optimum optical   最佳光学

parking lot   停车场

phase change material   相变材料

prior to   在……之前

Pyrenees   比利牛斯山脉(欧洲西南部)

Qinghai-Tibet Plateau   青藏高原

relay station   转播站;接力站

solar panel   太阳能电池板

the angle of incidence   入射角

the diffuse horizontal irradiance (DHI)   漫射水平辐照度

the global horizontal irradiance (GHI)   地球辐照度;辐射通量密度

the plane of orbit   轨道平面

vacuum collector tube   真空集热管

 Exercises

I . **Read the text and discuss over the following questions with your partner.**

1. What is the working principle of solar panels? Please describe it according to the relevant pictures in Text A.

2. What do you think is the application of solar energy, especially in China?

3. What is your opinion to avoid the disadvantages and try to make the best of solar energy?

II . **Fill in the blanks with the words and phrases in the text.**

The Sun's radiation reaches the Earth in a 1 _____ way because of its interaction with the atmosphere and the 2 _____ of incidence of sunrays. The angle of incidence varies according to two factors: the Earth's 3 _____ around its axis, which is very important for the alternation of day and night, and the 4 _____ of the Earth's axis as compared to the plane of its orbit, leading to a 5 _____ variation of the maximum height of the Sun on the horizon. On one hand, when the Sun is 6 _____ to the Earth's surface, the maximum concentration of sunrays on the ground is obtained. On the other hand, if the sunrays reach the Earth's 7 _____ with a certain inclination, the same amount of energy is 8 _____ over a larger surface. Therefore, solar energy can be highly exploited only within a belt included between 45° latitude south and north. For convenience and simplicity, the geographic distribution of total solar radiation on a global scale is divided in terms of 9 _____ into four broad belts around the Earth. These are described briefly with respect to the northern hemisphere, with the understanding that the same conditions apply to the 10 _____ belts in the southern hemisphere: the most favourable belt, moderately favourable belt, less favourable belt and least favourable belt.

Ⅲ. Please give the Chinese or English equivalents of the following terms.

1. solar panel
2. anti-reflective coating
3. artificial photosynthesis
4. backup power
5. beacon light
6. black light
7. COP
8. DC electrical power
9. electron flow
10. flat plate collector
11. 轨道平面
12. 保温箱;隔热箱
13. 停车场
14. 相变材料
15. 转播站;接力站
16. 入射角
17. 漫射水平辐照度
18. 真空集热管

19. 最佳光学

20. 多种用途

Ⅳ. Please translate the following sentences into Chinese or English.

1. Variation of solar energy resource is primarily caused by aerosol and cloud effects.

2. Passive solar techniques include orienting a building to the Sun, selecting materials with favorable thermal mass or light dispersing properties, and designing spaces that naturally circulate air.

3. The DC electricity generated by the solar panels is converted to AC current and can be used by all of our household appliances.

4. When the light hits the cell, a specific part of it gets absorbed by the material of the semiconductor.

5. All photovoltaic cells possess one or several electric fields that force the loosely flowing electrons to drift in a particular direction.

6. 通常将形成 N 型(负)硅的磷和形成 P 型(正)硅的硼的不纯混合物添加到纯硅中。

7. 尽管现有的建筑可以用类似的技术进行翻新,它们还是越来越多地作为主要或辅助的电源用于新建筑的建设中。

8. 这条位于北纬 15°和 35°之间的地带,是大自然赋予的太阳能应用条件最可观的区域。

9. 对于需适度供暖与制冷的气候,热泵提供了一种比火炉和空调更节能的选择。

10. 未来的科学家将面临这样的挑战,即制造常规尺寸的和更先进的太阳能电池板,以便在没有阳光的时候产生额外的能量。

Ⅴ. Please translate the following passage into Chinese.

Components can also form a solar power plant, which can be used for different power applications. Solar photovoltaic systems usually include batteries, storage, inverter, and control. The application of the building can be divided into stand-alone photovoltaic system and grid connected photovoltaic system. The application of solar cells provides lighting and other power for buildings. If solar energy can meet the energy demand of buildings, it can be called "zero energy house".

At present, most of the ways of utilizing solar energy in China are solar photovoltaic conversion, such as solar water heater, solar oven, solar room, solar drying, solar greenhouse, solar refrigeration and air conditioning, solar power generation and photovoltaic power generation system.

Tibet has the most abundant solar radiation energy in China. The annual total radiation in the whole region is mostly between 5000-8000 MJ/m^2, and the distribution is increasing from east to west.

Text B   Space-based Solar Power

1. Space-based Solar Power

Space-based solar power (SBSP) is the concept of collecting solar power in space (using an "SPS", that is, a "solar-power satellite" or a "satellite power system") for use on the Earth.

On the Earth, solar power is greatly reduced by night, cloud cover, atmosphere and seasonality. Some 30 percent of all incoming solar radiation never makes it to ground level. In space the Sun is always shining, the tilt of the Earth doesn't prevent the collection of power and there's no atmosphere to reduce the intensity of the Sun's rays. This makes putting solar panels into space a tempting possibility. Additionally, SBSP can be used to get reliable and clean energy to people in remote communities around the world, without relying on the traditional grid to a large local power plant.

SBSP differs from current solar collection methods in that the means used to collect energy would reside on an orbiting satellite instead of on Earth's surface. Some projected benefits of such a system are a higher collection rate and a longer collection period due to the lack of the diffusing atmosphere and night time in space.

2. How Does SBSP Work?

Self-assembling satellites are launched into space, along with reflectors and a microwave or laser power transmitter. Reflectors or **inflatable** mirrors spread over a vast **swath** of space, directing solar radiation onto solar panels. These panels convert solar power into either a microwave or a laser, and beam uninterrupted power down to the Earth. On the Earth, power-receiving stations collect the beam and add it to the electric grid.

The two most commonly discussed designs for SBSP are a large, deeper space **microwave transmitting satellite** and a smaller, nearer laser transmitting satellite. With the technology of microwave beaming, the rectenna receives the energy on the Earth.

新能源电力英语
English for Renewable Electricity Sources

2.1 Microwave transmitting satellites

Microwave transmitting satellites orbit the Earth in **geostationary Earth orbit** (GEO), about 35,000 km above the Earth's surface. Designs for microwave transmitting satellites are massive, with solar reflectors **spanning** up to 3 km and weighing over 80,000 metric tons. They would be capable of generating multiple gigawatts of power, enough to power a major US city.

The long wavelength of the microwave requires a long **antenna**, and allows power to be beamed through the Earth's atmosphere, rain or shine, at safe, low intensity levels hardly stronger than the midday Sun. Birds and planes wouldn't notice much of anything flying across their paths.

The estimated cost of launching, assembling and operating a microwave-equipped GEO satellite is in the tens of billions of dollars. It would likely require as many as 40 launches for all necessary materials to reach space. On the Earth, the **rectenna** used for collecting the microwave beam would be anywhere between 3 and 10 km in diameter, a huge area of land, and a challenge to purchase and develop.

2.2 Laser transmitting satellites

Laser transmitting satellites, orbit in **low Earth orbit** (LEO) at about 400 km above the Earth's surface. Weighing less than 10 metric tons, this satellite is a **fraction** of the weight of its microwave counterpart. This design is cheaper too; some predict that a laser-equipped SBSP satellite would cost nearly $500 million to launch and operate. It would be possible to launch the entire self-assembling satellite in a single rocket, drastically reducing the cost and time to production. Also, by using a laser transmitter, the beam will only be about 2 meters in diameter, instead of several kilometers, causing a drastic and important reduction.

To make this possible, the satellite's solar power beaming system employs a diode-pumped **alkali** laser. First demonstrated at LLNL in 2002 — and currently still under development there — this laser would be about the size of a kitchen table, and powerful enough to beam power to the Earth at an extremely high efficiency, over 50 percent.

While this satellite is far lighter, cheaper and easier to deploy than its microwave counterpart, serious challenges remain. The idea of high-powered lasers in space could

48

draw on fears of the militarization of space. This challenge could be remedied by limiting the direction that which the laser system could transmit its power. At its smaller size, there is a correspondingly lower capacity of about 1 to 10 megawatts per satellite. Therefore, this satellite would be best as part of a fleet of similar satellites, used together.

You could say SBSP is a long way off or pie in the sky (puns intended)— and you'd largely correct. But many technologies already exist to make this feasible, and many aren't far behind. While the Energy Department isn't currently developing any SBSP technologies specifically, many of the remaining technologies needed for SBSP could be developed independently in the years to come. And while we don't know the future of power harvested from space, we are excited to see ideas like this take flight.

2.3 Microwave beam

Sometimes it may happen that beam varies from its intended target. In the millisecond it takes for the target to discover that it's not receiving its signal, the beam shuts down. In other words, the microwave beam neither strays nor fries a nearby community. If the beam doesn't reach the target, it stops.

2.4 Rectennas

The receiving material itself is basically a wire mesh. In North America, the "footprint" is calculated to be an **ellipse** approximately one mile wide by one-and-one-half miles long. The mesh will be supported on poles 30 to 60 feet tall, much like the current power and telephone poles. One of the beauties of this system is that since it is wire mesh, it will not interfere with crops or animals under it. Rain and sunshine pass right through it. The microwave energy does not **penetrate** the mesh, so there's no danger there. In fact, once the novelty wears off, those near the rectenna will probably not even notice it's there. The rectenna will of course be fenced in and monitored, so the opportunity for **sabotage** or **vandalism** is no greater than any other power generation plant.

3. The Hurdles

But there are a few major hold-ups for this technology, namely, putting things into space. Satellites are incredibly expensive (50 million on the cheaper end), and though they may take less damage when in space, they could not simply be services when damage does occur. There's also a notable problem of how to transmit the energy back to the Earth. Solar panels on the ground convert photons into moving electrons and send them down wires to where they're needed, but we can't wire very well from space to the power plants. Ideas on multi-step processes involving photons becoming electrons becoming photons becoming electrons have been examined, but at each stage, energy would be lost.

Presently, the biggest obstacle to that vision is the expense of space travel. But recently the Power Sat Corporation announced the patent for "Space-Based Power Systems and Methods." The patent includes two technologies — Bright Star and Solar Powered Orbital Transfer (SPOT). These technologies will reduce launching and operation costs by roughly $1 billion for a 2,500 megawatt (MW) power system. Solar energy will be captured via solar power satellites (known as "power sat") and transmitted wirelessly via microwave to receiving stations at various points around the globe. The first technology, Bright Star, allows individual power sats to form a wireless power transmission beam without being physically connected to each other. This "electronic coupling" eliminates the need to handle large levels of power of the range of gigawatt in a single spacecraft. Because of Bright Star, one transmission beam may now come from hundreds of smaller power sats that effectively form one large satellite array. According to the CEO of Power Sat Corporation, William Maness, Bright Star will be on orbit for demonstration purposes in the 2017-2018, or at most within three years of that.

Besides the cost of implementing such a system, SBSP also introduces several new **hurdles**, primarily the problem of transmitting energy from orbit to the Earth's surface for use. Since wires extending from the Earth's surface to an orbiting satellite are neither practical nor feasible with current technology, SBSP designs generally include the use of some manner of wireless power transmission. The collecting satellite would convert solar energy into electrical energy on board, powering a **microwave transmitter** or **laser emitter**, and focus its beam toward a collector (rectenna) on the Earth's surface.

4. The History and Development of SBSP

In 1941, science fiction writer Isaac Asimov published the science fiction short story *Reason*, in which a space station transmits energy collected from the Sun to various planets using microwave beams.

The SBSP concept, originally known as Satellite Solar Power System (SSPS), was first described in November 1968. In 1973, Peter Glaser was granted the US patent number 3,781,647 for his method of transmitting power over long distances (e. g. from an SPSP to the Earth's surface) using microwaves from a very large antenna (up to one square kilometer) on the satellite to a much larger one, now known as a rectenna, on the ground.

Glaser then was a vice president at Arthur D. Little Inc. NASA signed a contract with ADL to lead four other companies in a broader study in 1974. They found that, while the concept had several major problems — chiefly the expense of putting the required materials in orbit and the lack of experience on projects of this scale in space — it showed enough promise to merit further investigation and research.

Between 1978 and 1981, the Congress authorized the Department of Energy (DOE) and NASA to jointly investigate the concept. They organized the Satellite Power System Concept Development and Evaluation Program. The study remained the most extensive performed to date (budget $ 50 million). Several reports were published investigating the engineering feasibility of such an engineering project, which included *Artist's concept of Solar Power Satellite in place.*

The project was not continued with the change in administrations after the 1980 US Federal elections. The Office of Technology Assessment concluded. Too little is currently known about the technical, economic, and environmental aspects of SPSP to make a sound decision whether to proceed with its development and deployment. In addition, without further research an SPSP demonstration or systems-engineering verification program would be a high-risk venture.

5. Further Development

In 2016, the US Naval Research Laboratory announced that Dr. Paul Jaffe, a spacecraft engineer, built a module to capture and transmit solar power. The idea is that a solar power satellite could transmit much cheaper electricity to the grid on the Earth. Jaffe refered to the module he created a "sandwich module. " This is how it works: the Sun sends out photons, and reflectors concentrate those photons on the sandwich module. The top, with the solar array, collects energy, and electronics in the middle turn it into a radio frequency. Antennas beam that to the Earth, and then they're converted back into electricity and sent to the grid.

In 2017, Chinese scientists announced they are building a solar power station in space, with an experimental one done by 2030 and a viable one completed by 2050. They said it could harness the Sun's energy almost all of the time and transmit it back down to the Earth. Of course, the biggest hurdles include making panels light and cheap enough to get up there, and then finding something to transport them to build it. Japan successfully tested a system that could transmit solar power from space to the Earth. Mitsubishi Heavy Industries tested the solar power system at Japan Space Systems, and saw 10 kilowatts were sent over microwaves to a receiver about 1,640 feet away, though the company never announced what percentage that was of the total power sent.

David Criswell suggests the Moon is the optimum location for solar power stations, and promotes lunar solar power. The main advantage he envisions is construction largely from locally available lunar materials, using **in-situ** resource utilization, with a teleoperated mobile factory and crane to assemble the microwave reflectors, and **rovers** to assemble and pave solar cells, which would significantly reduce launch costs compared to SBSP designs. Power relay satellites orbiting around the Earth and the Moon reflecting the microwave beam are also part of the project. A demo project of 1 GW starts at ＄50 billion.

Asteroid mining has also been seriously considered. A NASA design study evaluated a 10,000 ton mining vehicle (to be assembled in orbit) that would return a 500,000 ton asteroid fragment to geostationary orbit. Only about 3,000 tons of the mining ship that would be traditional aerospace-grade payload. The rest would be reaction mass for the mass-driver engine, which could be arranged to be the spent rocket stages used to launch the payload. Assuming that 100% of the returned asteroid was useful, and that the asteroid miner itself couldn't be reused, that represents nearly a 95% reduction in launch costs. However, the true merits of such a method would depend on a thorough mineral survey of the candidate asteroids.

Source:

1. http://en. wikipedia. org/wiki/In-situ_resource_utilization.

2. http://en. wikipedia. org/wiki/Asteroid_mining.

3. https://www. forbes. com/sites/lyndseygilpin/2015/09/25/space-based-solar-power-heres-how-it-could-work/#1f7cfe8947b4.

4. https://www. energy. gov/articles/space-based-solar-power.

5. https://www. mcgill. ca/oss/article/did-you-know-technology/space-based-solar-power.

New words and phrases

alkali	*adj.*	碱性的
	n.	碱
antenna	*n.*	天线;触角
asteroid	*n.*	小行星
ellipse	*n.*	椭圆
fraction	*n.*	小部分,少量,一点儿;分数
hurdle	*n.*	栏架;跨栏比赛;跨栏运动员
	v.	跨栏;克服
inflatable	*adj.*	需充气的;膨胀的
in-situ	*adj.*	原地的;现场的
penetrate	*v.*	穿过;进入;渗透,打入(组织、团体等);看透
rectenna	*n.*	硅整流二极管天线
rover	*n.*	漫游者;流浪者;巡视器
sabotage	*n.*	蓄意毁坏;捣乱;刻意阻碍
	v.	蓄意破坏(以防止敌方利用或表示抗议);妨碍;捣乱
span	*n.*	持续时间;范围;跨距,跨度
	v.	贯穿;涵盖(多项内容)
swath	*n.*	条,带;割下的一行草;(一镰刀的)刈幅
vandalism	*n.*	故意破坏公共财物罪;恣意毁坏他人财产罪
geostationary Earth orbit (GEO)		地球同步轨道
laser emitter		激光发射器
low Earth orbit (LEO)		近地轨道

microwave transmitter　微波发射器

microwave transmitting satellite　微波传输卫星

space-based solar power（SBSP）　空基太阳能；太空太阳能

 Exercises

Please think over the questions and come out with reasonable answers.

1. Is it plausible to generate solar power in space for use on the Earth?

2. What do you think are the greatest problems with SBSP?

3. Does it pay to try SBSP?

Keys to Exercises

Unit 3
Nuclear Power

Lead-in: *Nuclear power first came to vision as the powerful weapon during World War II. The explosion of nuclear bombs led to tremendous and lasting disaster to the environment and people's health. However, if used properly, nuclear power can be extremely beneficial to human beings. What is the usage of nuclear power? Do a brainstorm with your group members and list as many as you can. Figure 3-1 may serve as a clue.*

Fig. 3-1   The US nuclear powered ships

新能源电力英语
English for Renewable Electricity Sources

Text A   Nuclear Power

The text will focus on what nuclear power is, how nuclear power is generated, the application of nuclear power, the history of nuclear energy development, its world distribution, the advantages and disadvantages, the future of nuclear power and so on.

1. What Is Nuclear Power?

Nuclear energy (or atomic energy) is the energy released from the nucleus through nuclear reactions. Nuclear power is most frequently used in **steam turbines** to produce electricity in a nuclear power plant.

Nuclear power plants are the facilities responsible for converting the nuclear energy contained in the **uranium** atoms into electricity. The process to obtain this conversion is the result of a **thermodynamic** and mechanical process.

According to the International Energy Agency (IEA), in 2018, 11. 2 GW of additional nuclear capacity was connected to the grid, obtaining the largest increase since 1989. New projects were launched representing over 6 GW, and **refurbishment** projects are under way in many countries to ensure long-term operation of the existing fleet. Nevertheless, more efforts in terms of policies, financing and cost reductions are needed to maintain existing capacity and bring new **reactors** online. Under current trends, nuclear capacity in 2030 would amount to 497 GW, compared with 542 GW under the Sustainable Development **Scenario**.

2. How Is Nuclear Power Generated?

Nuclear power can be obtained from **nuclear fission**, **nuclear decay** and **nuclear fusion reactions**. Presently, the vast majority of electricity from nuclear power is produced by nuclear fission of uranium and **plutonium**. Nuclear decay processes are used in niche applications such as **radioisotope thermoelectric generators**.

Nuclear power can be generated through fission of **heavy elements**, such as uranium and the fusion of light elements, such as **deuterium**, **tritium**, and so on. The fission of heavy elements techniques has had practical application, and light elements in fusion technology is also being actively developed.

2.1   The fusion reaction

Fusion reactions (see Fig. 3-2) constitute the fundamental energy source of stars, including the Sun. The evolution of stars can be viewed as a passage through various stages as thermonuclear reactions and **nucleosynthesis** cause compositional changes over long time spans. Hydrogen (H) "burning" initiates the fusion energy source of stars and leads to the formation of **helium** (He). Generation of fusion energy for practical use also relies on fusion reactions between the lightest elements that burn to form helium.

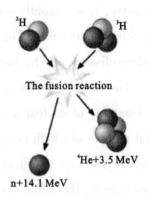

Fig. 3-2   **The fusion reaction**

In fact, the heavy isotopes of hydrogen — deuterium (D) and tritium (T) — react more efficiently with each other, and when they do undergo fusion, they yield more energy per reaction than do two hydrogen **nuclei**. As indicated in the picture, the hydrogen nucleus consists of a single **proton**; the deuterium nucleus has one proton and one neutron, while tritium has one proton and two neutrons.

2.2   The fission reaction

Nuclear fission is subdivision of a heavy atomic nucleus, such as that of uranium or plutonium, into two fragments of roughly equal mass. The process is accompanied by the release of a large amount of energy.

In nuclear fission, the nucleus of an atom breaks up into two lighter nuclei. The process may take place spontaneously in some cases or may be induced by the excitation of the nucleus with a variety of particles (e. g. neutrons, protons, deuterons, or **alpha particles**) or with electromagnetic radiation in the form of **gamma rays**. In the fission process, a large quantity of energy is released, radioactive products are formed, and several neutrons are emitted. These neutrons can induce fission in a nearby nucleus of fissionable material and release more neutrons that can repeat the sequence, causing a chain reaction in which a large number of nuclei undergo fission and an enormous amount of energy is released. If controlled in a nuclear reactor, such a chain reaction can provide power for society's benefit. If uncontrolled, as in the case of the so-called atomic bomb, it can lead to an explosion of awesome destructive force.

Although the early experiments involved the fission of ordinary uranium with slow neutrons, it was rapidly established that the rare isotope uranium-235 was responsible for this phenomenon. The more abundant isotope uranium-238 could be made to undergo fission only by fast neutrons with energy exceeding 1 MeV.

The fission process may be best understood through a consideration of the structure and stability of nuclear matter. Nuclei consist of nucleons (neutrons and protons), the total number of which is equal to the mass number of the nucleus. The actual mass of a nucleus is always less than the sum of the masses of the free neutrons and protons that constitute it, the difference being the mass equivalent of the energy of formation of the nucleus from its constituents. The conversion of mass to energy follows Einstein's equation, $E = mc^2$, where "E" is the energy equivalent of a **mass**, "m", and "c" is the velocity of light. This difference is known as the mass defect and is a measure of the total **binding energy** (and, hence, the stability) of the nucleus. This binding energy is released during the formation of a nucleus from its constituent nucleons and would have to be supplied to the nucleus to decompose it into its individual nucleon components.

2.3 Nuclear decay

When an atom decays, it can do so via one of three natural methods. These methods are known as alpha, beta, and gamma. In alpha decay, the unstable nucleus ejects an alpha particle, which is composed of two neutrons and two protons. Another way of stating this is that the nucleus decays by ejecting a helium-4 nucleus. In **beta decay**, the nucleus ejects a beta particle, which is either an electron (beta minus) or a positron (beta plus)[1]*. At first glance, this would seem to be wrong, as a nucleus is comprised of protons and neutrons and contains no electrons or positrons. But they are produced in the nucleus whenever a neutron decays into a proton, an electron, and a neutrino or a proton decays into a neutron, a positron, and a neutrino. The last way of decaying is via gamma decay, which is when electromagnetic radiation is given off by the nucleus as the protons and neutrons become more tightly bound.

There is another way for a nucleus to decay, though this method usually involves the actions of humans. A nucleus can be forced to break apart if it is hit by particles from outside of the nucleus.

3　The Application of Nuclear Power

Nuclear power is widely used in the world. It is used in many fields. Apart from nuclear weapons, the most important and known use of nuclear energy is the generation of electricity. Moreover, it is used in industry, agriculture, hydrology, mining and so on.

The generation of electricity: after the Second World War, the main use of nuclear energy was the generation of electric power. Nuclear power today makes a significant contribution to electricity generation, providing 10% of global electricity supply in 2018. In advanced economies, nuclear power accounts for 18% of generation and is the largest low-carbon source of electricity. However, its share of global electricity supply has been declining in recent years. That has been driven by advanced economies, where **nuclear fleets** are ageing, additions of new capacity have **dwindled** to a trickle, and some plants built in the 1970s and 1980s have been retired. This has slowed the transition towards a clean electricity system. Despite the impressive growth of solar and wind power, the overall share of clean energy sources in total electricity supply in 2018, at 36%, was the same as it was 20 years earlier because of the decline in nuclear. Halting that slide will be vital to stepping up the pace of the decarbonisation of electricity supply.

The use in industry: nuclear power is used in the development and improvement of processes, for measurements, automation and quality control. It is used as a **prerequisite** for the complete automation of high-speed production lines, and is applied to process research, mixing, maintenance and the study of wear and corrosion of facilities and machinery. Nuclear technology is also used in the manufacture of plastics and in the sterilization of single-use products.

The use in agriculture: nuclear technology is very useful in the control of insect pests, in the maximum use of water resources, in the improvement of crop varieties or in the establishment of the necessary conditions to optimize the effectiveness of fertilizers and water.

Other uses of nuclear technology occur in **disciplines** such as hydrology, mining or the space industry.

4. The History of Nuclear Power Development

In 1932, physicist Ernest Rutherford discovered that when lithium atoms were "split" by protons from a proton accelerator, immense amounts of energy were released in accordance with the principle of mass-energy equivalence. However, he and other nuclear physics pioneers Niels Bohr and Albert Einstein believed harnessing the power of the atom for practical purposes anytime in the near future was unlikely, with Rutherford labeling such expectations "moonshine."

In 1932, James Chadwick discovered the neutron, which was immediately recognized as a potential tool for nuclear experimentation because of its lack of an electric charge. Experimentation with **bombardment** of materials with neutrons led Frédéric and Irène Joliot-Curie to discover induced radioactivity in 1934. Further work by Enrico Fermi in the 1930s focused on using slow neutrons to increase the effectiveness of induced radioactivity. Experiments bombarding uranium with neutrons led Fermi to believe he had created a new, **transuranic** element, which was **dubbed hesperium**.

But in 1938, German chemists Otto Hahn and Fritz Strassmann, along with Austrian physicist Lise Meitner and Meitner's nephew, the Austrian-British physicist Otto Robert Frisch, conducted experiments with the products of neutron-bombarded uranium, as a means of further investigating Fermi's claims. They determined that the relatively tiny neutron split the nucleus of the massive uranium atoms into two roughly equal pieces, contradicting Fermi. This was an extremely surprising result: all other forms of nuclear decay involved only small changes to the mass of the nucleus, whereas this process — dubbed "fission" as a reference to biology — involved a complete **rupture** of the nucleus. Numerous scientists, including Leó Szilárd, who was one of the first, recognized that if fission reactions released additional neutrons, a self-sustaining nuclear chain reaction could result. Once this was experimentally confirmed and announced by Frédéric Joliot-Curie in 1939, scientists in many countries (including the United States, the United Kingdom, France, Germany, and the Soviet Union) petitioned their governments for support of nuclear fission research, just on the **cusp** of World War II, for the development of a nuclear weapon.

During the late 1940s and early 1950s, research programs in the United States, United Kingdom, and the Soviet Union began to yield a better understanding of nuclear fusion, and investigators embarked on ways of exploiting the process for practical energy production. Fusion reactor research focused primarily on using magnetic fields and electromagnetic forces to contain the extremely hot **plasmas** needed for thermonuclear fusion.

Plasma physics theory in the 1950s was incapable of describing the behaviour of the plasmas in many of the early magnetic confinement systems. However, the undeniable potential benefits of practical fusion energy led to an increasing call for international cooperation. American, British, and Soviet Union's fusion programs were strictly classified until 1958, when most of their research programs were made public at the Second Geneva Conference on the Peaceful Uses of Atomic Energy, sponsored by the United Nations. Since that time, fusion research has been characterized by international collaboration. In addition, scientists have also continued to study and measure fusion reactions between the lighter elements so as to arrive at a more accurate determination of reaction rates. The formulas developed by nuclear physicists for predicting the rate of fusion energy generation have been adopted by astrophysicists to derive new information about the structure and evolution of stars.

Work on the other major approach to fusion energy, **inertial confinement fusion** (ICF), was begun in the early 1960s. The initial idea was proposed in 1961, only a year after the reported invention of the laser, in a then-classified proposal to employ large pulses of laser energy (which no one then quite knew how to achieve) to implode and shock-heat matter to temperatures at which nuclear fusion would proceed vigorously. Aspects of inertial confinement fusion were declassified in the 1970s and, especially, in the early 1990s to reveal important aspects of the design of the targets containing fusion fuels. Very painstaking and sophisticated work to design and develop short-pulse, high-power lasers and suitable millimetre-sized targets continues, and significant progress has been made.

5. Distribution of Nuclear Power in the World

There are nuclear power plants in operation in **circa** 30 countries worldwide. A majority of these countries are located in Europe, North America, or Asia (especially East Asia and South Asia). In 2012, the year after the Fukushima Daiichi nuclear disaster, the global nuclear electricity generation was at its lowest level since 1999.

Only five countries in the world derive a majority of their electricity from nuclear power plants. They are France, Slovakia, Ukraine, Belgium, and Hungary.

The single largest producer of electricity from nuclear power plants is the United States, where well over 8 million MWh was produced in 2017.

According to IEA, nuclear energy supplies 20 percent of the US electricity needs, less than the 31.7 percent that comes from natural gas and the 30.1 percent from coal, and only slightly more than the 17.1 percent provided by renewables such as hydropower, wind and solar. But some countries depend on the atom more heavily. France, for example, gets 72 percent of its electricity from nuclear plants, and Sweden gets about 40 percent from them, according to a report from April 2018.

The country where nuclear power is growing fastest right now is China, where over 25 reactors are under construction. Other notable countries are India, Russia, and South Korea, where nuclear power generation is also expanding at a fast pace.

Examples of countries that are going in the opposite direction, i. e. , planning to phase out their nuclear power plants, are Germany, Switzerland, Belgium, and Spain. Italy closed all its nuclear power stations by 1990 after a national **referendum**. Sweden and the Netherlands are also considering a phase-out, but the political situation there is more fractured and no commitment has been made to firm plans.

In the 21st century, some modern nuclear power plants were built to provide power to the grid as a new source of energy and nuclear power was applied to fields like space industry with the development of theories on nuclear power.

6. The Advantages and Disadvantages of Nuclear Power

Nuclear power has avoided about 55 Gt of CO_2 emissions over the past 50 years, nearly equal to 2 years of global energy-related CO_2 emissions.

The use of nuclear energy has a low greenhouse emission. According to **the International Atomic Energy Agency (IAEA)**, nuclear power is one of the lowest emitters of greenhouse gases available to generate electricity, and makes a significant contribution to reducing greenhouse gas emissions worldwide. Nuclear energy contributes more effectively than other techniques to solve problems such as " the greenhouse effect" and acid rain. So it is environmentally friendly. In this way, the use of nuclear energy facilitates the sustainable development of the world.

In addition, the use of nuclear energy facilitates the development of economy. Nuclear energy produces cheap electricity, so we are not afraid of energy shortage. As of 2018, the International Atomic Energy Agency knew of 449 operable civilian fission reactors in the world used to generate electricity for the power grid. Their combined capacity was 394 GW. IAEA also listed 58 reactors that were under construction and 154 planned reactors. The 58 reactors had a combined planned capacity for 63 GW, while the 154 planned reactors were intended to generate a combined 157 GW.

Despite the contribution from nuclear and the rapid growth in renewables, energy-related CO_2 emissions hit a record high in 2018 as growth of electricity demand outpaced increases in low-carbon power. As the technology is not well developed, human beings still have problems in controlling some of the disadvantages brought by nuclear power.

Nuclear power plants have high start-up costs. Plants must invest heavily in containment systems and emergency plans. Extensive backup systems must be built and **contingency** plans must be developed to handle the rare threat of core meltdown[2]*.

Radioactive material following such as a damaging earthquake, can have severe consequences for the environment. A nuclear power plant accident is capable of releasing dangerous radiation that harms human beings and the environment. Even though the Nuclear Regulatory Commission monitors plant operation and construction closely, nuclear mishaps are still possible and have occurred. Historical accidents have left great damages to the world. The Three Mile Island nuclear reactor in Pennsylvania experienced a partial meltdown in 1979. In 1986, a reactor in Chernobyl sent radioactive material as far as Sweden and large swathes of the surrounding region are still considered uninhabitable today. More recently, three reactor building explosions and three core meltdowns occurred at Japan's Fukushima nuclear plant after an earthquake and tsunami rocked the country in 2011. The accident contaminated air, water, homes and farms and displaced 160,000 people. In 2015, extremely low levels of radiation from the Fukushima mishap were recorded on North American shores.

7. Future Development of Nuclear Power

The future of nuclear power varies greatly between countries, depending on government policies. Some countries, many of them in Europe, such as Germany,

Belgium, and Lithuania, have adopted policies of nuclear power phase-out. At the same time, some Asian countries, such as China, South Korea, and India, have committed to rapid expansion of nuclear power. Many other countries, such as the United Kingdom and the United States, have policies in between. Japan was a major generator of nuclear power before the Fukushima accident, but as of August 2016, Japan has restarted only three of its nuclear plants, and the extent to which it will resume its nuclear program is uncertain.

Countries envisaging a future role for nuclear account for the bulk of global energy demand and CO_2 emissions. But to achieve a **trajectory** consistent with sustainability targets — including international climate goals — the expansion of clean electricity would need to be three times faster than that at present. It would require 85% of global electricity to come from clean sources by 2040, compared with just 36% today. Along with massive investments in efficiency and renewables, the trajectory would need an 80% increase in global nuclear power production by 2040.

Source：

1. https://www. iea. org/fuels-and-technologies/nuclear.

2. https://www. britannica. com/science/nuclear-fission.

3. https://core. ac. uk/reader/5222143.

4. http://www. iccf11. org/.

5. https://www. bp. com/content/dam/bp/business-sites/en/global/corporate/pdfs/energy-economics/statistical-review/bp-stats-review-2019-full-report. pdf.

New words and phrases

bombardment　*n.*　炮轰；轰炸

circa　*prep.*　大约

contingency　*n.*　偶然性；意外事故

cusp　*n.*　尖头；尖端

deuterium　*n.*　重氢；氘

discipline　*n.*　学科

dub　*v.*　给……起绰号；授予……称号；结账

dwindle　*v.*　减少,变小,缩小;衰落,变坏,退化

helium　*n.*　氦;氦气

hesperium　*n.*　钚(94 号元素)

mass　*n.*　物质

meltdown　*n.*　熔毁(指核燃料过热,熔化反应堆活性区或外罩);灾难性崩溃(或瓦解)

niche　*n.*　壁龛;(~market)有利可图的市场(或形势等)

nuclei　*n.*　原子核(nucleus 的复数形式)

nucleosynthesis　*n.*　核合成

plasma　*n.*　等离子体

plutonium　*n.*　钚(94 号元素)

prerequisite　*n.*　前提

proton　*n.*　质子

radioisotope　*n.*　放射性同位素

reactor　*n.*　反应器;(核)反应堆

referendum　*n.*　公民投票;请示书

refurbishment　*n.*　整修

rupture　*n.*　断裂,破裂

scenario　*n.*　方案;设想,可能发生的情况;剧本

thermodynamic　*adj.*　热动力学的

thermoelectric　*adj.*　热电的

trajectory　*n.*　(物)轨道,轨线;(军)弹道

transuranic　*adj.*　超铀的

tritium　*n.*　超重氢;氚

uranium　*n.*　铀

alpha particle　α粒子

beta decay　β衰变

binding energy　结合能;束缚能

gamma rays　伽马射线

heavy element　重元素

inertial confinement fusion (ICF)　惯性约束聚变

nuclear decay　核衰变

nuclear fission　核裂变

nuclear fleet　核舰队

nuclear fusion reaction　核聚变反应

radioisotope thermoelectric generator　放射性同位素热电发生器

steam turbine　蒸汽涡轮机

the International Atomic Energy Agency（IAEA）　国际原子能机构

Notes

1* In beta decay, the nucleus ejects a beta particle, which is either an electron (beta minus) or a positron (beta plus). 在β衰变中,原子核弹出一个β粒子,它要么是电子(β-衰变),要么是正电子(β+衰变)。

2* A meltdown occurs when a reactor core overheats and radioactive fuel escapes. If that hot fuel melts through barriers designed to keep it in, radioactive material could escape into the area outside the reactor. Safety measures have tightened since the Three Mile Island incident. 当反应堆堆芯过热而放射性燃料逸出时,就会发生熔毁。如果热燃料将外罩熔化,放射性物质就可能散逸至反应堆外部。自美国三里岛核事故以来,相应的安全措施已经加强。

Exercises

Ⅰ. Read the text and discuss over the following questions with your partner.

1. What is the working process of nuclear fusion? Please describe it according to the picture of nuclear fusion in Text A.

2. What do you think is the future of nuclear power, especially in China?

3. How to avoid the adverse impacts of nuclear power?

Ⅱ. Fill in the blanks with the words and phrases in the text.

The fission process may be best understood through a consideration of the structure and stability of 1 _____ matter. Nuclei consist of nucleons (neutrons and protons), the total 2 _____ of which is equal to the mass number of the 3 _____. The actual mass of a nucleus is always less than the sum of the masses of the free neutrons and 4 _____ that constitute it, the difference being the mass equivalent of the energy of formation of the nucleus from its 5 _____. The 6 _____ of mass to energy follows Einstein's 7 _____, $E = mc^2$, where "E" is the energy equivalent of a mass, "m", and "c" is the velocity of light. This difference is known as the mass defect and is a measure of the total 8 _____ energy (and, hence, the stability) of the nucleus. This binding energy is 9 _____ during the formation of a nucleus from its constituent nucleons and would have to be supplied to the nucleus to 10 _____ it into its individual nucleon components.

III. Please give the Chinese or English equivalents of the following terms.

1. nuclear fusion

2. nuclear fission

3. nuclear reactor

4. heavy element

5. radioisotope thermoelectric generator

6. 氘

7. 核衰变

8. 氚

9. (核)反应堆熔毁

10. 蒸汽涡轮机

IV. Please translate the following sentences into English.

1. 据目前趋势,到 2030 年的核电容量将达到 497 吉瓦,而在可持续发展方案下则为 542 吉瓦。

2. 核电站是负责将铀原子中所含的核能转化为电能的地方。

3. 目前,核电增长最快的国家是中国,在那里,超过 25 座核反应堆正在建设中。

4. 核能发电是温室气体排放量最低的发电方式之一,为减少全球温室气体排放做出了重要贡献。

5. 科学家们也继续研究和测量较轻元素之间的聚变反应,以便更准确地确定反应速率。

6. 日本的福岛核电站事故造成大气、水、家园和农场的污染,并迫使 16 万人撤离。

V. Please translate the following passage into Chinese.

The future of nuclear power varies greatly between countries, depending on government policies. Some countries, many of them in Europe, such as Germany, Belgium, and Lithuania, have adopted policies of nuclear power phase-out. At the same time, some Asian countries, such as China, South Korea, and India, have committed to rapid expansion of nuclear power. Many other countries, such as the United Kingdom and the United States, have policies in between. Japan was a major generator of nuclear power before the Fukushima accident, but as of August 2016, Japan has restarted only three of its nuclear plants, and the extent to which it will resume its nuclear program is uncertain.

Countries envisaging a future role for nuclear account for the bulk of global energy demand and CO_2 emissions. But to achieve a trajectory consistent with sustainable development targets — including international climate goals — the expansion of clean electricity would need to be three times faster than that at present. It would require 85% of global electricity to come from clean sources by 2040, compared with just 36% today. Along with massive investments in efficiency and renewables, the trajectory would need an 80% increase in global nuclear power production by 2040.

Text B　Nuclear Accidents and Radioactive Waste

1. Nuclear Accidents

The risk and consequence of nuclear accidents is a heavily debated topic. Various technical measures and work routines have been adopted to reduce the risk of accidents and minimize the negative impact when an accident does occur, but accidents are still happening and some of them even have serious effects.

1.1　Examples of nuclear plant accidents

The Chernobyl disaster (April 26, 1986).

This disaster occurred in Chernobyl, Kiev Oblast, Ukrainian SSR, the Soviet Union. Fuel rods in reactor 4 of the Chernobyl power plant overheated, which led to an explosion and a meltdown. Radioactive material was dispersed over large parts of Europe. 30 direct deaths.

The Fukushima Daiichi disaster (March 11, 2011).

This disaster occurred at the Fukushima Daiichi nuclear power plant in Okuma, Japan. A tsunami damaged the plant's five active reactors, and loss of electrical power led to overheating and meltdowns. 2 direct deaths.

The Kyshtym disaster (September 29, 1957).

This disaster occurred at the nuclear fuel reprocessing plant Mayak in Ozyorsk, a closed city located near Kyshtym in Russia, the Soviet Union. The cooling system in a tank containing more than 70 tons of liquid radioactive waste failed. The temperature rose, resulting in evaporation, and eventually an explosion of dried waste. The explosion released an estimated 800 PBq of radioactivity.

The Sellafield fire (October 10, 1957).

A fire at the British atomic bomb project in Sellafield, Cumberland, the United Kingdom destroyed the core and released an estimated 740 TBq of iodine-131. Fortunately, a **rudimentary** smoke filter on the main chimney prevent the leak from becoming bigger. It is therefore not considered a disaster.

The accident at the Lucens reactor (January 21, 1969).

A loss-of-coolant accident occurred at the experimental Lucens reactor in Vaud,

Switzerland, and this resulted in a partial core meltdown. Fortunately, the cavern — which had been heavily contaminated with radioactive material — was sealed quickly, which prevented the incident from becoming a disaster.

The Three Mile Island accident (March 28, 1979).

A loss-of-coolant accident occurred at reactor 2 of the Three Mile Island nuclear generating station near Harrisburg in Pennsylvania, the United States, and resulted in a partial core meltdown. In total, approximately 93 PBq of radioactive gases, and approximately 560 GBq of iodine-131 was released into the environment[1]*.

1. 2 Examples of nuclear-powered submarines involved in nuclear accidents

American submarine SSN-593 in 1963. The USS Thresher (SSN-593) was lost at sea on 10 April 1963 during its first deep dive test (DDT) after a nine-month post-shakedown availability (PSA) at Portsmouth Naval Shipyard (PNSY). It was the first and remains the world's worst nuclear submarine disaster, killing all 129 people on board.

Soviet submarine K-140 in 1968.

Soviet submarine K-222 in 1980.

Soviet submarine K-431 in 1985. In the K-431 nuclear submarine accident, ten sailors were killed and 49 others suffered radiation damage when the submarine exploded while refueling it. The incident took place on August 10, 1985.

1. 3 Examples of accidents involving radiotherapy

The 1990 **Clinic of Zaragoza radiotherapy accident in Spain.**

The 1996 **radiotherapy accident at San Juan de Dios Hospital in San José, Costa Rica.**

2. Radioactive Waste

Radioactive waste, especially high-level waste, is hazardous to living organisms. The radioactivity in the waste products will decrease over time, but for some types of waste it will take a very long time before the waste stops being dangerous. Radioactive waste should therefore be isolated and confined to prevent living organisms from being exposed to it.

How long radioactive waste remains dangerous depends largely on the **half-life** of the element. **Cesium-137** and **strontium-90** have a half-life of roughly 30 years, which means that after 30 years, roughly 50% of the atoms of the radioactive material will have disintegrated.

For plutonium, twenty radioactive isotopes have been characterized, and their half-life vary dramatically. When discussing long-term management of plutonium isotopes, it is therefore very important to know exactly which isotopes we are talking about. Plutonium-241 does for instance have a half-life of less than 15 years, while it is 88 years for plutonium-238. Plutonium-244 has a half-life that exceeds 80 million years, the half-life for plutonium-242 is more than 373,000 years, and the half-life for plutonium-239 is 24, 110 years.

2.1   High-level radioactive waste

Nuclear reactors produce high-level radioactive waste (HLW). One example of HLW are used fuel rods that have been removed from the core. They contain both fission products and transuranic elements, are highly radioactive and tend to be hot.

An average 1,000-MW nuclear power plant will produce around 27 tonnes of spent nuclear fuel in a year. The waste from spent fuel rods will consist chiefly of cesium-137 and strontium-90, but can also contain plutonium.

Two examples of long-lived fission products from nuclear reactors that are of special concern are **Technetium-99** and **Iodine-129**. After a few thousand years, these two will be the dominating radioactive materials left in the spent fuel rods created today. The half-life for technetium-99 is 220,000 years and the half-life for Iodine-129 is 15.7 million years.

2.2   Reprocessing or recycling

Reprocessing or recycling of spent nuclear fuel is possible with current technology. It still generates waste, and is therefore not a total solution, but it can be used to reduce the quantity of waste that must be put into long-term storage. Many programs for reprocessing or recycling spent nuclear fuel are currently active worldwide.

2.3   Methods of disposal

Many different methods of long-term disposal of radioactive waste (especially HLW) have been discussed. Right now, the dominating suggestions are deep geological

burial in a mine or in a **borehole**.

So far, there is no repository available for civilian HLW anywhere in the world; all radioactive waste that is being stored is stored in storage facilities intended for short term or medium term storage. One long-term repository is currently being constructed in western Finland, near the municipality Eurajoki. This is a deep geological repository named Onkalo, excavated in granit bedrock. It is expected to start receiving Finnish HLW in 2020.

In the United States, HLW from several sources is stored in The Morris Operation in Grundy, Illinois, but this storage isn't intended for long-term keeping of HLW. In this facility, spent nuclear fuel assemblies are kept in a spent fuel storage pool. HLW is also stored at 70 different nuclear power plant sites throughout the United States, since there is no long-term storage facility available to send it to. **The Yucca Mountain Nuclear Waste Repository**, approved in 2002, was intended to be a deep geological repository for HLW within the Yucca Mountain, but federal funding for the project was stopped in 2011 under the Obama Administration.

2.4 Various suggested methods of disposal of radioactive waste

Deep borehole disposal is the concept of disposing HLW in extremely deep boreholes. No country has implemented this, but some experiments have been carried out in the United States. (That program has now been closed due to lack of funding.)

Ocean disposal. This method was implemented by several countries, but there are now international agreements in place that ban this practise. Examples of countries that practised ocean disposal of HLW before the ban are the United States, the Soviet Union / Russia, the United Kingdom, Japan, Switzerland, Germany, France, Italy, the Netherlands, South Korea, Belgium, and Sweden.

Sub seabed disposal. This method is never implemented and it is now banned by international agreements.

Direct injection. This method has been used by the United States and the Soviet Union.

Disposal in outer space. Currently, this method is considered too expensive and has not been implemented by any country.

Long term above ground storage. This method has not been implemented by any country, since it's considered unsafe for long-term storage.

Rock melting. This method has not been implemented by any country.

Disposal at **subduction zones**. This method has not been implemented by any country.

Disposal in ice sheets. This would violate the Antarctic Treaty.

Source：

1. http://www.iccf11.org/nuclear-accidents.

2. http://www.iccf11.org/radioactive-waste.

New words and phrases

borehole   *n.*   钻孔

rudimentary   *adj.*   初级的；初步的；起码的

cesium-137   *n.*   铯137

strontium-90   *n.*   锶90

Technetium-99   *n.*   锝99

Iodine-129   *n.*   碘129

half life   半衰期

sub seabed disposal   亚海床处置

subduction zone   俯冲带

the Chernobyl disaster   切尔诺贝利灾难

the Fukushima Daiichi disaster   福岛第一核电站灾难

the Kyshtym disaster   克什特姆事故

the Sellafield fire   塞拉菲尔德大火

the accident at the Lucens reactor   卢肯斯反应堆事故

the Three Mile Island accident   三里岛事故

the Clinic of Zaragoza radiotherapy accident in Spain   西班牙萨拉戈萨诊所放射治疗事故

the radiotherapy accident at San Juan de Dios Hospital in San José, Costa Rica   哥斯达黎加圣何塞圣胡安·迪奥斯医院的放射治疗事故

the Yucca Mountain Nuclear Waste Repository   尤卡山核废料储存库

Notes

1 * In total, approximately 93 PBq of radioactive gases, and approximately 560 GBq of iodine-131 was released into the environment[1*]. 总共释放到环境中约 93 PBq 的放射性气体和约 560 GBq 的碘 131。PBq——放射性活度单位。放射性元素每秒有一个原子发生衰变时，其放射性活度即为 1 贝可。$P = 10^{15}$, $G = 10^9$。

Exercises

Please discuss over the following questions with your teammates after reading the text. You can support your view with more information from online or other channels.

1. Tell your teammates one of the nuclear disasters that you know and give examples of its impacts on human life.

2. Since nuclear power produces such serious radioactive waste, why some countries still exploit nuclear power and build more nuclear power plants?

Keys to Exercises

3. Can you bring up a practical and reasonable strategy of disposal method of radioactive waste after learning the many disposal methods in the text?

Unit 4
Water Energy

Lead-in : *Water is essential to our life. The function of water in our life can not be exaggerated. Human beings can find no other planet to dwell on because of the lack of water on the other planet. People began to use water for different purposes since ancient times. Please work in groups to discuss the usages of water through brainstorming. Figure 4-1 may serve as a clue.*

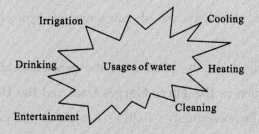

Irrigation   Cooling

Drinking   Usages of water   Heating

Entertainment   Cleaning

Fig. 4-1 Brain-storming: usages of water

Text A   Hydroelectricity

To learn about the modern usage of water energy, we will focus on hydroelectricity. When water is used for generating electricity, we have a special term for it — **hydroelectricity**, which plays a key role in renewable energy to improve the

sustainable development. The United Nations Sustainable Development Goals emphasized to meet the global food security challenges by mechanized farming; access of clean water challenges by renewable freshwater **withdrawal**s; clean energy issues determined by clean fuel and cleaner technologies; and **combat** climate change by limiting **anthropogenic** emissions of carbon, fossil fuel, and greenhouse gas emissions in the air. As is mentioned, water plays an important role in sustainable development goals of the United Nations. Hydroelectricity is reliable so long as there is enough water on the earth. Furthermore, the generation of hydroelectricity does little harm to our environment. The text will mainly focus on three aspects of hydroelectricity, i. e. , what is hydroelectricity, how it is generated and why people choose hydroelectricity; meanwhile, some relevant issues will be introduced, such as the history of hydroelectricity, world distribution, problems with hydroelectricity, and so on.

1. What Is Hydroelectricity?

Hydroelectricity is the term referring to electricity generated by hydropower, the production of electrical power through the use of the gravitational force of falling or flowing water. It is the most widely used form of renewable energy. Hydropower is used primarily to generate electricity. Broad categories of hydroelectricity include the followings.

Conventional hydroelectricity: it refers to the generation of hydroelectricity with hydroelectric dams, such as **the Three Gorges Dam** and **the Hoover Dam**.

Run-of-the-river hydroelectricity (ROR): it captures the kinetic energy in rivers or streams, without a large **reservoir** and sometimes without the use of dams.

Small hydro project: the generating capacity of it is usually 10 megawatts at most and often has no artificial reservoirs.

Micro hydro project: it provides a few kilowatts or up to a few hundred kilowatts to isolated homes, villages, or small industries.

Conduit hydroelectricity project: it utilizes water which has already been diverted for use elsewhere; in a municipal water system, for example.

Pumped-storage hydroelectricity: it stores energy in the form of potential energy of water, pumped from a lower elevation reservoir to a higher elevation.

Pressure buffering hydropower: it uses natural sources (waves, for example) for water pumping to turbines while exceeding water is pumped uphill into reservoirs and releases when incoming water flow isn't enough.

2. How Is Hydroelectricity Generated?

Aside from a plant for electricity production, a hydropower facility consists of a water reservoir enclosed by a dam, which captures energy from the movement of a river. The gates in the dam can open or close depending on how much water is needed to produce a particular amount of electricity. Reservoirs can be used for recreation, wildlife **sanctuaries**, and sources of drinking water. Figure 4-2 illustrates the process of hydroelectricity generation.

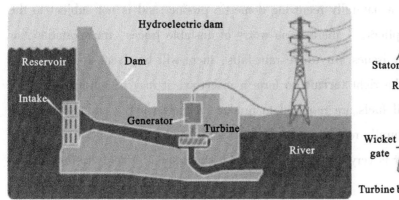
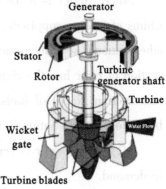

(a) Flowing water drives a water wheel or turbine

(b) A generator attached to the turbine produces electricity

Fig. 4-2　Hydroelectricity generation

http://www. eoearth. org/view/article/161501/.

Hydroelectric power derives from transfers of kinetic energy when a volume of water flows down from a high elevation source, such as a reservoir, through a generator turbine and out to the stream-bed below.

Water flows down a pipe called a **penstock**, through a water turbine and drives the turbine blades to rotate. The turbine drives an electrical generator through the shaft which turns and makes the generator work to generate electricity. The energy extracted depends on water volume of the reservoir and the height differential. The water finally flows downstream. The whole process makes no waste of water.

Water use in power plants has two components: withdrawal and consumption. Water withdrawal is the act of removing water from a local water source; the withdrawn

water may or may not get returned to its source or be made available for use elsewhere. Water consumption is the use of water in a power plant in a way such that the water is not returned, usually because it is lost to evaporation.

3. Why Do People Develop Hydroelectricity?

It is clear that water energy provides an alternative to new sources of energy. The application of hydroelectricity brings more advantages than disadvantages. Here list a few.

(1) No pollution is released into the atmosphere and no waste that requires special containment is produced.

Since "water is a naturally recurring domestic product and is not subject to the whims of foreign suppliers, " there is no worry of unstable prices, transportation, or other national security issues. Anywhere rain falls, there will be rivers. If a particular section of river has the right **terrain** to form a reservoir, it may be suitable for dam construction. No fossil fuels are required to produce the electricity, and the Earth's **hydrologic cycle** naturally replenishes the "fuel" supply.

(2) Hydropower is very convenient because it can respond quickly to fluctuations in demand.

A dam's gates can be opened or closed on command, depending on daily use or gradual economic growth in the community. The production of hydroelectricity is often slowed in the nighttime when people use less energy. When a facility is functioning, no water is wasted or released in an altered state; it simply returns unharmed to continue the hydrologic cycle.

(3) Hydroelectric power is also very efficient and inexpensive.

Modern hydro turbines can convert as much as 90% of the available energy into electricity. The best fossil fuel plants are only about 50% efficient. In the US, hydropower is produced for an average of 0. 7 cents per kWh. This is about one-third the cost of using fossil fuel or nuclear and one-sixth the cost of using natural gas, as long as the costs for removing the dam and the silt it traps are not included.

(4) Hydropower has become the leading source of renewable energy. It provides more than 97% of all electricity generated by renewable sources worldwide. Other sources including solar, geothermal, wind, and biomass account for less than 3% of

renewable electricity production. In the US, 81% of the electricity produced by renewable sources comes from hydropower. Worldwide, about 20% of all electricity is generated by hydropower.

(5) Hydropower is environmentally friendly as it reduced CO_2 emissions. The use of hydropower avoids the burning of 22 billion gallons of oil or 120 million tons of coal each year.

(6) Compared with other methods of power generation, hydroelectricity eliminates the **flue gas emissions** from fossil fuel combustion, including pollutants such as sulfur dioxide, **nitric oxide**, carbon monoxide, dust, and mercury in the coal. Hydroelectricity also avoids the hazards of coal mining and the indirect health effects of coal emissions. Compared with nuclear power, hydroelectricity generates no nuclear waste, and has none of the dangers associated with uranium mining, nor nuclear leaks. Compared with wind farms, hydroelectricity power stations have a more predictable load factor. If the project has a storage reservoir, it can generate power when needed. Hydroelectric stations can be easily regulated to follow variations in power demand.

However, hydroelectricity also has some disadvantages. They mainly include the following.

(1) The high cost of hydroelectric facilities.

Like all power plants, hydroelectric plants are very expensive to build, and must be built to a very high standard. Since the most feasible sites for dams are in hilly or mountainous areas, the faults that often created the topography pose a great danger to the dams and therefore the land below them for thousands of years after they have become useless for generating power.

(2) The building of dams for hydroelectric power can cause a lot of water access problems.

The creation of a dam in one location may mean that those down river no longer have control of water flow. This can create controversy in places where neighboring countries share a water supply.

(3) Generating electricity requires substantial amounts of water.

Not any place that has water is suitable for building a dam. Only places that can provide enough amount of water are considered significant to built a hydroelectricity plant.

(4) Electricity can also affect water quality.

Producing electricity can have significant implications for water quality. For example, water used to cool electricity-generating steam exits the power plant at substantially higher temperatures — up to 18 °F hotter at power plants in summer. Minerals unearthed during fuel mining and drilling can contaminate groundwater, which in turn affects drinking water and local ecosystems. Coal mining and combustion create wastes with dangerous toxins such as **mercury**, lead, and **arsenic**; and improper storage or disposal of those wastes can contaminate water supplies.

4. The History of Hydroelectricity

Hydropower has been used since ancient times to grind flour and perform other tasks. In the mid-1770s, French engineer Bernard Forest de Bélidor published *Architecture Hydraulique* which described **vertical- and horizontal-axis** hydraulic machines.

Man-made waterfalls dams were constructed throughout the 1900s in order to maximize this source of energy. By the 1940s, the best sites for large dams had been developed. At the beginning of the 20th century, many small hydroelectric power plants were being constructed by commercial companies in mountains near metropolitan areas. Grenoble, France held the International Exhibition of Hydropower and Tourism with over one million visitors. By 1920, as 40% of the power produced in the United States was hydroelectric, **the Federal Power Act** was enacted into law. The Act created the Federal Power Commission to regulate hydroelectric power plants on federal land and water. As the power plants became larger, their associated dams developed additional purposes including flood control, irrigation and navigation. Federal funding became necessary for large-scale development and federally owned corporations, such as the **Tennessee Valley Authority** (1933) and the **Bonneville Power Administration** (1937) were created. Additionally, the **Bureau of Reclamation** which had begun a series of western US irrigation projects in the early 20th century was now constructing large hydroelectric projects such as the 1928 Hoover Dam. **The US Army Corps of Engineers** were also involved in hydroelectric development, completing the Bonneville Dam in 1937 and being recognized by the Flood Control Act of 1936 as the premier federal flood control agency.

Hydroelectric power plants continued to become larger throughout the 20th century. Hydropower was referred to as white coal for its power and plenty. Hoover Dam's initial 1,345 MW power plant was the world's largest hydroelectric power plant in 1936; it was eclipsed by the 6,809 MW **Grand Coulee Dam** in 1942. **The Itaipu Dam** opened in South America as the largest in 1984, producing 14,000 MW but was surpassed in 2008 by **the Three Gorges Dam** in China at 22,500 MW. Hydroelectricity would eventually supply some countries, including Norway, the Democratic Republic of the Congo, Paraguay and Brazil, with over 85% of their electricity. The United States currently has over 2,000 hydroelectric power plants that supply 6.4% of its total electrical production output, which is 49% of its renewable electricity.

"Over the past 100 years, the United States has led the world in dam building." Secretary of the Interior Bruce Babbitt recently observed that, "on average, we have constructed one dam every day since the signing of the Declaration of Independence." Of the 75,187 dams in the US, less than 3% are used to produce 10%-12% of the nation's electricity. The US owns over 2,000 facilities of hydroelectricity.

5. The World Distribution of Hydroelectricity

How much of the world's electricity is supplied by hydroelectric power?

China (8 mtoe) and Brazil (4 mtoe) posted the largest contributions.[1] European generation rebounded by 9.8% (12.9 mtoe), almost offsetting its steep decline in the previous year. Pacific's global share has increased significantly in recent years: in 2018, Asia Pacific accounted for 41% of global consumption. 20 years ago it accounted for only 20%.

The ranking of hydro-electric capacity is either by actual annual energy production or by **installed capacity power rating**. A hydro-electric plant rarely operates at its full power rating over a full year; the ratio between annual average power and installed capacity rating is the capacity factor. The installed capacity is the sum of all generator nameplate power ratings.

This is an indication of the total electricity supply by hydroelectric power in several different countries:

— 99% in Norway;

— 75% in New Zealand;

— 50% in developing countries;

— 25% in China;

— 13% in the US.

6. Current Problems in Hydroelectricity Generation

As with other forms of economic activity, hydropower projects can have both a positive and a negative environmental and social impact, because the construction of a dam and power plant, along with the **impounding** of a reservoir, creates certain social and physical changes. Hydropower projects can also have indirect consequences, contributing to global warming: reservoirs accumulate plant material, which then decomposes, emitting methane in uneven bursts.

From 2013 to 2015, a series of projects were carried out to deal with hydroelectric problems. One of the projects was water loss control. **San Diego Gas & Electric**, Southern California Edison and Southern California Gas conducted leak detection for participating water utilities, computed the energy intensity of saved water, estimated energy saved through water loss control, evaluated **efficacy** of **acoustic leak detection** vs. other technologies and estimated cost of repairing leaks vs. cost of energy saved. Another project dealt with water sector over-generation and flexible demand response. The project was a pilot study exploring opportunities for the State's Water Sector to provide electric reliability through Flexible Demand Response using water storage and pumping undertaken by Southern California Edison. Another project dealt with the issue of accelerating drought resilience by exploring innovative strategies and technologies for building drought resilience while also supporting electric reliability and reducing GHG (greenhouse gas) emissions.

In 2019, the results of Tapio **elasticity** found that carbon-fossil-greenhouse gas emissions' contamination in water-energy-food's resources are quite visible that exhibit weak decoupling state, expensive negative decoupling state, and strong decoupling state in the different decade's data, which substantiate the ecological cost in water-energy-food's resources. [2]* The results emphasized the need to adopt different sustainable instruments in a way to limit carbon-fossil-greenhouse gas emissions in water-energy-food resources through cleaner production technologies, renewable energy

82

mix, environmental certification, anti-dumping tariff duty, strict environmental regulations, and etc. These instruments would be helpful to achieve environmental sustainability agenda for mutual exclusive global gains.

7. Future Development of Hydroelectricity

Hydro technology has now entered the era of rapid development. Based on the previous experience, some standards of hydroelectricity have been issued. People can avoid some of the defects of previous work and have a series of criteria to follow. In other words, hydroelectricity has come along a scientific track.

According to a research article, as to the renewable energy in G-7 countries[3*], empirical findings indicate that increasing biomass energy consumption was efficient to reduce carbon emission in France, Germany, Japan and the United States; increasing hydroelectricity usage was efficient to reduce carbon emission in Italy and the United Kingdom; wind energy consumption reduced emission in Canada and solar energy usage was efficient on reducing emission in France and Italy for observed period. Moreover, in case of **panel**, it is found that increasing hydroelectricity, biomass and wind energy consumption reduced carbon emissions while the impact of solar energy consumption is statistically insignificant in G-7 countries. In addition, the hydroelectricity consumption was found the most efficient renewable energy source to reduce environmental pollution for the panel of G-7 countries.

According to another research article, water being one of the cheapest and renewable sources of energy, is being used to produce one-quarter of the total electricity production in Pakistan. The forecasted values of hydroelectricity consumption revealed an average annual increment of 1. 65% with a **cumulative** increase of 23. 4% up to the year 2030. The results were compared with the hydroelectricity generation plans of the Government of Pakistan for its effectiveness.

Another research article proposed regional water scarcity issue used water footprint (WF). Hydropower is a crucial no-fossil energy source, but it may cause environmental damages by huge water consumption mainly from evaporation. Facing great hydroelectric capacity and water scarcity issue in China, water loss from hydropower deserves further valuation. The results of the study illustrated a significant regional variation. The average gross and net WF were at 3. 021 (range of 0. 08-122. 31) L/kWh and 0. 0763 (range of 0-9. 638) L/kWh, respectively. WFs also showed a considerable seasonal variation with apparent regional characters. Although hydropower WF was relatively low in China, several provinces with water scarcity issue were inappropriate for hydropower development.

From the several research articles above, it can be generalized that hydroelectricity is and will play an important role in the future development of different countries in the world. There is still such issue as scarcity and seasonal change of water which may hinder the development of hydroelectricity. It remains a heavy task for human beings to adjust to local development by applying hydroelectricity appropriately.

Source:

1. https://kns. cnki. net/KCMS/detail/detail. aspx? dbcode = SJES&dbname = SJESTEMP _ U&filename = SJESBCDDCD073BEA1CDEE3BA0B45B6CFDE21&v.

2. http://en. wikipedia. org/wiki/Grenoble.

3. https://www. bp. com/en/global/corporate/energy-economics/statistical-review-of-world-energy/hydroelectricity. html.

 New words and phrases

anthropogenic   *adj.*   人类起源论的;人为的

arsenic   *n.*   砷;砒霜

combat   *n.*   格斗;搏斗;战斗

conduit   *n.*   导管

cumulative   *adj.*   累积的

efficacy   *n.*   功效

elasticity   *n.*   弹性,弹力;灵活性

hydroelectricity   *n.*   水力发电

impound   *v.*   搁置;保留;蓄水

mercury   *n.*   汞;水银

panel   *n.*   镶板;小组;讨论组;专家组

penstock   *n.*   水道

reservoir   *n.*   水库

sanctuary   *n.*   避难所;庇护

terrain   *n.*   地形,地势;领域;地带

withdrawal   *n.*   移开;撤回,撤开;收回,取回

acoustic leak detection   声漏检测

Architecture Hydraulique   《水利建筑》

Bonneville Power Administration   (美)邦纳维尔电力管理局

Bureau of Reclamation   农垦局

flue gas emission   烟气排放

hydrologic cycle   水循环

installed capacity power rating   额定装机容量

nitric oxide   氧化亚氮

San Diego Gas & Electric   圣地亚哥天然气电力公司

vertical- and horizontal-axis   垂直和水平轴

the Federal Power Act   联邦电力法案

Tennessee Valley Authority   (美)田纳西州流域管理局

Grand Coulee Dam   大古力水坝

the Hoover Dam   胡佛大坝

The US Army Corps of Engineers   美国陆军工兵部队

The Itaipu Dam   伊泰普大坝

the Three Gorges Dam   三峡大坝

Notes

1* China (8 mtoe) and Brazil (4 mtoe) posted the largest contributions.   中国(800 万桶油当量)和巴西(400 万桶油当量)贡献最大。

2* In 2019, the results of Tapio elasticity found that carbon-fossil-greenhouse gas emissions' contamination in water-energy-food's resources are quite visible that exhibit weak decoupling state, expensive negative decoupling state, and strong decoupling state in the different decade's data, which substantiate the ecological cost in water-energy-food's resources.   2019 年,Tapio 弹性结果发现,碳化石温室气体排

放对水、能源及食品资源的污染相当明显,在不同十年的数据中表现出弱去耦状态、高成本的负去耦状态和强去耦状态,证实了水、能源及食品资源的生态成本。

3* G-7: The panel of G-7 countries includes seven countries — Canada, France, Germany, Italy, Japan, the United States, and the United Kingdom. 七国集团国家小组包括加拿大、法国、德国、意大利、日本、美国及英国七个成员国。

Exercises

Ⅰ. Read the text and discuss over the following questions with your partner.

1. What is the working process of hydroelectricity? Please describe it according to the picture of hydroelectricity generation in Text A.

2. What do you think is the future of hydroelectricity, especially in China?

3. How to avoid adverse impacts of hydro dams and reservoirs?

Ⅱ. Fill in the blanks with the words and phrases in the text.

Hydroelectric power derives from transfers of 1 _____ energy when a volume of water flows down from a high 2 _____ source, such as a reservoir, through a 3 _____ turbine and out to the stream-bed below.

Water flows down a pipe called a 4 _____, through a water turbine and drives the turbine 5 _____ to rotate. The turbine drives an electrical generator through the 6 _____ which turns and makes the generator 7 _____ to generate electricity. The energy extracted depends on water 8 _____ of the reservoir and the 9 _____ differential. The water finally flows 10 _____. The whole process makes no waste of water.

Ⅲ. Please give the Chinese or English equivalents of the following terms.

1. 声漏检测
2. 水足迹
3. 水循环
4. 垂直和水平轴

5. 额定装机容量

6. the average annual increment

7. conduit hydroelectricity project

8. pumped-storage hydroelectricity

9. dangerous toxin

10. white coal

IV. Please translate the following sentences into English.

1. 水力发电指利用水能发电,利用落水或流水的重力产电。

2. 发电厂用水形式有两种:调水和耗水。

3. 水力发电非常方便,因为它能对需求的波动做出快速反应。

4. 煤炭开采和燃烧会产生含有汞、铅和砷等危险毒素的废物,对这些废物的不当储存或处置会污染水源。

5. 随着发电厂的规模越来越大,与之相关的水坝也发展出防洪和灌溉的额外用途。

V. Please translate the following passage into Chinese.

From 2013 to 2015, a series of projects were carried out to deal with hydroelectric problems. One of the projects was water loss control. San Diego Gas & Electric, Southern California Edison and Southern California Gas conducted leak detection for participating water utilities, computed the energy intensity of saved water, estimated energy saved through water loss control, evaluated efficacy of acoustic leak detection vs. other technologies and estimated cost of repairing leaks vs. cost of energy saved. Another project dealt with water sector over-generation and flexible demand response. The project was a pilot study exploring opportunities for the State's Water Sector to provide electric reliability through Flexible Demand Response using water storage and pumping undertaken by Southern California Edison. Another project dealt with the issue of accelerating drought resilience by exploring innovative strategies and technologies for building drought resilience while also supporting electric reliability and reducing GHG (greenhouse gas) emissions.

Text B   Current Status and Future Prospects of Hydroelectricity Standardization

1. Introduction

Standards are the foundations of production, trade and service. They guarantee the construction of projects and the quality of products. Now, standards have become the important measures and approaches of improving the competitiveness of the nation.

With the rapid progress in the electricity industry of China, the corresponding standardization work has also achieved considerable progress, of which the hydroelectricity standards have laid a solid technological foundation for the development of hydroelectricity.

So far there are 1,489 valid electricity standards, which include 234 national standards and 1,225 industrial standards. 91 of them are related to the category of hydroelectricity, including hydroelectricity design, hydro-turbine, hydro-generator, **electrical apparatus**, hydropower plant automation and metal structure, which almost covers all the fields in hydroelectricity, and can fulfill the requirements from production, construction, operation and management and promote the technological advancement in hydroelectricity technology.

2. Current Status of Hydroelectricity Standardization

2.1   A system of Standardization Technical Committees established

The Standardization Technical Committee (**STC**), as the supervisory institution in the specific field, plays an important role in maintaining the technological level of standards and **implementing** standardization in **corresponding fields**.

Today, the STC in hydroelectricity consists of five technical committees — planning and designing, hydro-turbine, hydro-generator and electrical equipment, hydropower plant automation, as well as metal structure and **hoist** of hydropower plant. The STC for hydroelectricity planning and designing is in charge of the standardization on hydroelectricity exploration, planning, hydraulic structure design, **electromechanical design**, reservoir **emigration**, and computer application in hydroelectricity planning, and etc. The STC for hydro-turbine is responsible for carrying out the standardization on the technology, operation and installation of hydro-turbines. The STC for hydro-generator and electrical equipment is in charge of the standardization on related technology, operation and installation. The STC for hydropower plant automation takes charge of the standardization on corresponding technology, operation and field tests, involving **excitation**, governor, monitor, water status and automation elements. The STC for metal structure and hoist of hydropower plant implements standardization in related technology, operation, and tests. Besides, the STC for hydro construction is in charge of implementing standardization in organizing the construction design of **permanent** and **temporary** projects of major projects, building materials, construction technology (including installation and assembly), construction quality and safety, as well as the equipment, machines and instruments developed and produced by the construction enterprises.

According to Management Rules for Standardization Technical Committee of Electricity Industry, members of STCs shall represent varieties, including engineers and managers from production, manufacture, design, construction, research and development, teaching and supervision, and also young members who have good technical background and are interested in standardization work. All the work of STCs has been carried out in accordance with corresponding laws and regulations.

2.2　Hydroelectricity standardization

Standards in design, construction and operation.

With the development of Chinese electricity industry speeded up, the technological progress and innovation in electricity industry is also accelerated. Until now, the unit capacity of hydro-turbines has reached as high as 700 MW. Extra large-scale hydropower bases and **unattended hydropower plants** have come into being and become popular. Hydro technology has now entered the era of rapid development. A series of important standards, such as DL/T5186 Specifications on Electromechanical Design of Hydropower Plants, DL/T1006 Guidelines on Maintenance for Equipment in

Hydropower Plants, and DL/T5208 Guidelines on Design of Pumped-Storage Hydropower Stations, have been established to meet the requirements from hydroelectricity construction.

Standards in security, environmental protection and energy conservation.

A series of corresponding standards, such as Specifications on Fireproofing in Hydraulic Structure and Hydroelectricity Design and Specifications on Labor Safety and Health in Hydraulic Structure and Hydroelectricity Design, have been listed in working schedule. Some of them are even under **draft**.

Standards in disaster-relief.

After the occurrence of ice storm in 2008, the capability of **disaster relief** for electrical equipment has been paid high attention to by related departments. According to the requirements of the National Energy Administration, the Standardization Scheme on Enhancing Anti-Disaster Ability of Power Systems was established, in which it is clearly stated to improve the disaster-relief level of electrical facilities, and enhance the supervision on implementing corresponding **compulsory** standards.

Standards in quality assessment and acceptance.

To improve the project quality and do a better job in inspection, assessment and acceptance of the project quality, a series of standards such as Standards on Quality Rating Evaluation of Basic Construction Units in Hydraulic Structures and Hydroelectricity have been put into the task list for the next few years to enhance the management of headstream control, **tache** control and process control. Some of them have been completed, and others are still being formulated.

Standards in project construction translated into English version.

In recent years, Chinese hydroelectricity enterprises (involving exploration, design, construction and investment, and etc.) have participated actively in international competition, and have achieved remarkable market share. They contracted to build oversea projects, which promotes the export of electromechanical products, architectural materials and labor services. However, there are very few standards in our country written in English, especially for the construction standards. Therefore, the Chinese standards are not widely accepted by foreign countries. The standards of a third-party have to be employed, which will **adversely** affect our market competition and technology export.

In order to adapt to international competition environments after China's entry into WTO, encourage the Chinese enterprises to participate in the construction of international projects and implement the development strategy called "going abroad", and furthermore, **propagate** our standards to the world, corresponding institutions and STCs have translated the current standards in project construction into English version since 2005, and established a system of standards in English version step by step. After two-year work, 82 of these standards have been planned to be translated into English version. Among them, 33 have been approved, and 28 have been published.

Published standards reviewed to extend effectiveness.

According to the requirements for standardization, certain standards were reviewed by STCs and institutions which had participated in drafting these standards every year. Those that have been published for more than 5 years must be reviewed to determine whether to be continued, updated or **abolished**.

3. Future Work of Hydroelectricity Standardization

The standardization work will focus on the construction of pumped-storage hydropower stations, the **vibration** of large hydro-generator units, the condition-based maintenance of large hydro-generators, and large metal structures and hoists. Some important standards will be established, such as **specifications** of large hydro-generators, specifications of computer monitoring system design for pumped-storage hydropower stations, rules for the optimal operation of **cascading** hydropower plants, criteria for the online monitoring system and condition-based maintenance of large hydro-generators, rules for the design of large **ship lifters**, large **ship locks** and metal structures, and so on.

Source:

https://www.ixueshu.com/download/cbdda1caac6215c230432f9051a4015b318947a18e7f9386.html.

New words and phrases

abolish   *v.*   废除

adversely   *adv.*   不利地

cascade   *v.*   喷流;瀑布状物;串联

compulsory   *adj.*   必修的;必须的

draft   *v.*   起草

      n.   草案

emigration   *n.*   移民

excitation   *n.*   激发;激励

hoist   *v.*   提升;举起;增强

      n.   起重机;举起

implement   *v.*   执行;实行

permanent   *adj.*   永久的;常驻的

propagate   *v.*   宣传;传播

specification   *n.*   技术指标;详细说明

tache   *n.*   环;扣

temporary   *adj.*   临时的

vibration   *n.*   振动

corresponding field   对应领域

disaster relief   赈灾;救灾

electrical apparatus   电器设备

electromechanical design   机电设计

ship lifter   船舶起重器

ship lock   船闸

the Standardization Technical Committee（STC） 标准化技术委员会(简称 STC)

unattended hydropower plant   无人值守型水电站

Exercises

Please answer the following questions.

1. What is the significance of standards?

2. Are the standards drafted by China all approved by the international projects?

3. What shall members of STCs represent according to Management Rules for Standardization Technical Committee of Electricity Industry?

Keys to Exercises

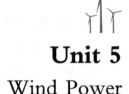

Unit 5
Wind Power

Lead-in: *Wind power has been used by human beings in many ways for thousands of years. Ancient mariners used sails to capture the wind and explore the world. Farmers once used windmills to grind their grains and pump water. Nowadays, more and more countries are using wind turbines to wring electricity from the breeze. Over the past decade, the use of wind turbine has increased more than 25 percent a year. China is one of the countries which are increasingly developing wind power. On the top of many village hills, wind turbines are standing high and spinning constantly. Talk with your teammates the wind plants you know and share with them your point of view on the advantages and disadvantages of wind power (see Fig. 5-1).*

Fig. 5-1 Wind power

Text A　Wind Power

1. Introduction

Following the invention of the electric generator, engineers began harnessing wind energy to produce electricity. Wind power generation succeeded in the United Kingdom and the United States in 1887-1888. However, modern wind power is said to have started in Denmark, where horizontal-axis wind turbines were built in Askov in 1891, and a 22.8-meter wind turbine for electric generation started operation in 1897. Since then, wind generation has spread from Europe and the United States to the world. Most new wind power projects have turbine capacities of around 2 MW onshore and 3-5 MW offshore.

Global installed wind generation capacity (including both onshore and offshore capacity) has increased nearly 50 times in the past two decades, from 7.5 GW in 1997 to more than 371 GW in 2014 (WWEA, 2015a). Denmark, a leader in wind power generation, installed the world's first offshore wind farm, consisting of 11 wind turbines of 0.45 MW each, in 1991 (Carbon Trust, 2008). By the end of 2014, the total installed offshore wind capacity worldwide was 8.8 GW (GWEC, 2015).

2. Wind Technologies and Performance

Wind power and basic elements of wind turbines.

Wind, the movement of the air, is the result of temperature differences in different places. Uneven heating results in a difference in atmospheric pressure, which causes the air to move. The **kinetic** energy of the moving air (or wind) is transformed into electrical energy by wind turbines or wind energy conversion systems. The wind forces the turbine's **rotor** to spin, changing the kinetic energy to **rotational** energy by moving a shaft which is connected to a generator, thereby producing electrical energy through electromagnetism.

Wind power **is proportional to** the dimensions of the rotor and to the cube of the wind speed. Theoretically, when the wind speed doubles, the wind power increases eight times. The main factors of the output power are the swept area (related directly to the length of the blades) and the wind speed. Over time, the size of wind turbines has increased continually. In 1985, turbines had a rated capacity of 0. 05 MW and a rotor diameter of 15 metres (EWEA, 2011).

The largest commercially available wind turbines to date reach 8. 0 MW each, with a rotor diameter of 164 metres. The average capacity of newly installed wind turbines has increased from 1. 6 MW in 2009 to 2. 0 MW in 2014 (Broehl, Labastida and Hamilton, 2015).

Wind power systems are categorised primarily by the grid connection (connected / stand-alone), installation characteristic (onshore / offshore) and wind turbine type (vertical / horizontal-axis). The specific system configuration is determined mainly by the wind condition (especially wind speed), land availability (or where the plant is sited), grid availability, turbine size and height, and blade size.

For vertical-axis turbines, used primarily for small generation capacities, the axis of rotation is vertical to the wind flow / ground. The turbines are independent of wind direction, and some can generate electricity at both low wind speeds and low noise levels, making them particularly suited to urban areas. In addition, heavy components, such as the generator, can be mounted at ground level. This results in easier maintenance and lighter-weight towers and is expected to contribute to the stability of floating foundations for offshore wind. However, vertical-axis turbines are less efficient at turning wind energy into mechanical power, and some require a starting device. Moreover, it is difficult to control the rotation speed. Horizontal-axis turbines are being used commercially all over the world. The rotating shaft is mounted parallel to the wind flow / ground, and the turbines can have two types of rotors: up-wind and down-wind. The advantage of up-wind turbines is that they are hardly influenced by the turbulence caused by the tower. However, a **yaw mechanism** is needed to align the turbine with the prevailing wind. Meanwhile, down-wind rotors for large turbines are emerging in Japan. Because they can sufficiently catch winds blowing upwards, they can be a promising technology for improving the stability and safety of floating offshore wind facilities.

The basic elements of the wind-power system are the blades, the rotor hub, the rotor shaft, the nacelle, the rotor brake, the gearbox, the generator and controller, the tower and the transformer.

The blades capture and convert the wind's energy to rotational energy. The number of blades also influences the structure and ability of wind turbines. Typical large turbines are up-wind, horizontal-axis turbines with three rotor blades. Because of the better balance of gyroscopic forces, most modern wind turbines use three rotor blades; fewer blades would mean slower rotation, requiring more from the gear box and transmission. Modern blades are typically made from reinforced fibreglass and are shaped **aerodynamically**, similar to the profile of aircraft wings. Although carbon fibre-reinforced plastic is a stronger material, the cost remains high. Smaller blades can be made from (laminated) wood, which has strength and weight advantages.

The rotor hub transfers the rotational energy to the rotor shaft, which is fixed to the rotor hub. The other end of the rotor shaft is connected to the gearbox, which changes the low rotating speed from the blades to a high rotating speed for input to the generator. Direct-drive systems without a gearbox are also available, and their market share is growing in Europe and China. Advantages of gearless turbines include their compact structure, lower risk of breakdown and simpler maintenance.

Wind turbines incorporate a control system to prevent excessive rotation speeds in high wind, which could otherwise break the blades or other components. The two methods for controlling the speed of the blade, or delivering the power output from the blade to the rotating shaft, are pitch control and stall control. A pitch control system actively adjusts the angle of the blades to the wind speed. The rotor hub includes a pitch mechanism, and the control system features a brake. A stall protection system decreases the rotational speed by using the aerodynamic effects of the blades when the wind speed is too high, lowering the efficiency to protect the turbine from damage.

The high-speed rotating shaft connected to the gearbox forces the shaft of the generator to rotate, converting the rotational energy to electricity through the use of electromagnetic induction. Typically, two types of generators are used with wind turbines: induction (or **asynchronous**) generators, which usually require excitation power from the network; and **synchronous** generators, which can start in isolation and

produce power corresponding directly to rotor speed. The rotor shaft, rotor brake, gearbox and generator components are housed within a nacelle, which is directly connected to the blades at a high elevation and is one of the main structures of the wind generating system. To rotate the nacelle so as to align the wind turbine with the direction of wind, the wind turbine has a mechanism called a yaw system. Modern large systems are installed with an active yawing system which is controlled by an electric control system with a wind direction sensor.

The rotor blades, rotor hub and nacelle are supported by and elevated on a tower. The height of the tower is determined by the rotor diameter and the wind conditions of the site. Many towers are made from steel tubes which allow access to the nacelle inside the tower, even in bad weather. Newer tower types include the "space frame tower", which improves the logistics of installation and transport (GE, 2014). In large wind turbines, the towers contain electric cables, a ladder or lift for maintenance, and occasionally a control system. The control system has three main functions: controlling turbines (e.g. the rotating speed and yaw direction), monitoring and collecting operational data (e.g. on weather conditions, or output / input data for the system, including electricity voltage and current, rotating speed, yaw direction, vibration frequency of blade and nacelle) and communicating with operators.

A transformer is usually placed at ground level and transforms the electricity from the generator to the required voltage on the grid. The aerodynamic loss of energy at the blade / rotor is about 50%-60%, the mechanical loss of energy at the gear is 4% and the electromechanical loss at the generator is 6%. Overall, the efficiency of wind power generation is 30%-40% (NEDO, 2013).

3. Onshore and Offshore Wind Farms

In general, a power generating facility which contains a number of wind turbines is called a "wind farm". The basic elements of the wind farm are wind turbines, monitoring facilities, substations and transmission cables. If they are offshore, wind farms also need port facilities for maintenance.

Recently, particularly in Europe, offshore wind farms have gained higher market shares as a result of supportive government policies, inspired by the idea that offshore wind is faster and more stable than onshore wind. However, the capital and maintenance costs of offshore wind farms are several times higher than of onshore wind farms.

The noticeable difference between onshore and offshore wind farms is the foundation. An onshore wind turbine stands on a concrete foundation, whereas offshore turbines have their foundations in the water (floating) or on the sea bed (fixed-bottom). Fixed-bottom foundations can have varying types of structures.

(1) Monopole, jacket, tripod, gravity-based and suction bucket. The monopole structure is the simplest and hence the most common, but can only be used in shallow water (up to 30 metres in depth) (IEA, 2013).

(2) Floating foundations are typically used at depths exceeding 50-60 metres, because the cost of fixed-bottom foundations becomes prohibitive in deeper waters. Floating structures are currently in the demonstration phase.

(3) Offshore wind farms are designed to withstand elements of the severe marine environment, such as waves and seawater, and have additional operational requirements, such as access to the turbine. Efforts to reduce maintenance costs therefore are critical.

4. Small Wind Turbines

There is no internationally agreed upon definition of small wind turbines. However, the International Electrotechnical Commission (IEC) defines a small wind turbine as having a rotor swept area of less than 200 square metres, equating to a rated power of some 50 kW. Other national organisations in major small wind markets such as China, the US and the UK define a small wind turbine as having a rated power of less than 100 kW. Until around 1980, most wind turbines had a capacity of less than 100 kW.

Small wind turbines generally require a higher capital cost per kW and have a lower efficiency (load factor) compared to big wind farms. However, small wind turbines are beneficial for electricity storage or off-grid electric supply in rural / grid-isolated areas where expensive diesel generators are being used. Small wind turbines have been used mainly for off-grid electricity generation and water pumping in isolated areas.

The horizontal-axis small wind turbine is a proven technology which usually uses

permanent magnet generators and direct-drive technology. It continues to dominate the small wind turbine market. In 2011, 74% of manufacturers supplied the horizontal-axis turbines, 18% supplied vertical-axis turbines and 6% tried to provide both types. The average horizontal-axis model was estimated at 10.8 kW in capacity, while the average vertical-axis model was 7.4 kW (WWEA, 2013). Innovative vertical-axis turbine designs are being used mainly in urban environment, particularly in China (Global Data, 2015).

A major challenge in installing a small wind turbine is assessing the wind resource. The resource assessment process is similar to that for larger turbines; however, it is expensive due to the high cost of management tools and long-term measurement efforts. Generally, small wind turbines are installed as stand-alone units, not as wind farms, which dramatically increases the planning costs per unit installed. The height of the tower is also a key factor for small wind turbines. To reduce the negative effects of turbulence caused by surrounding obstacles, a taller tower is better; however, it has a higher cost. Most small wind turbines are below 30 metres in height. Further innovation to reduce costs and to improve the efficiency of turbines at lower heights are key challenges for the development of small wind technology.

As of the end of 2013, more than 755 MW of small wind power capacity was installed worldwide, with 41% of the facilities sited in China, 30% in the US and 15% in the UK (WWEA, 2015b).

5. Insights for Policy Makers

Wind, or the kinetic energy of air flow, has been used in transport, industry and agriculture for thousands of years. The rise of modern wind turbines, which harness this energy and turn it into electricity, is a story of scientific and engineering skill coupled with strong **entrepreneurial spirit**. Wind power continues to expand worldwide, reflecting the reduced cost of turbines, expanding policy support and growing investor recognition of the positive characteristics of wind generation. In 2014, wind power reached a more than 3% share of the world's electricity supply. In 2015, China led this development with capacity additions of 32.9 GW, followed by the United States (8.6 GW) and Germany (4.9 GW). By the end of 2015, more than 434 GW of wind power capacity had been installed worldwide (WWEA, 2016).

Efforts are being made to improve the economic efficiency of wind power facilities. Wind farms are being built to maximise energy production and to minimise capital and operating costs, while remaining within the constraints **imposed** by the site. Once the site constraints are defined, "micro-siting" is performed to **optimise** the layout design. For most wind power projects, the economics depend far more on the **fluctuating** costs of energy production than on infrastructure costs. For both onshore and offshore facilities, the dominant parameter for layout design is the maximisation of energy production (as opposed to, for example, whether turbines are located close to one another for ease of maintenance or grid connection).

The recent development of large onshore wind farms has reduced the number of remaining sites with good wind resource potential, especially in more densely populated areas of Europe. Some European countries are developing offshore wind power by taking advantage of the relatively shallow seabed adjoining the continent. Because wind speeds at sea are generally higher than those on land, and there are fewer obstacles at sea which can cause turbulence, offshore wind power is more efficient than onshore wind power. However, because offshore wind is an emerging sector and faces unique challenges related to working at sea, it has higher construction and operation costs and hence a higher overall generation cost. Nevertheless, offshore wind generation is expanding. By the end of 2014, cumulative global offshore wind capacity was approximately 8. 8 GW (GWEC, 2015).

Like most renewable energy sources, wind power is capital-intensive, and reductions in capital costs are important for realizing wind energy projects. Although wind operations have no fuel cost, reducing the operation and maintenance (O&M) costs is key to improving the economics of wind power. Some countries have introduced financial supports such as **feed-in tariffs** to secure greater income and to reduce investor risk. The levelized cost of electricity (LCOE) for typical wind farms in 2014 was in the range of USD 0. 06-0. 10 per kilowatt-hour (kWh) for onshore wind to USD 0. 12-0. 21 per kWh for offshore wind. The best wind projects in the world are consistently providing electricity for USD 0. 05 per kWh, without financial support (IRENA, 2015a). Small wind turbines (with a rated power of less than 50-100 kilowatts (kW), as defined by the International Electrotechnical Commission and some countries) use mature technology and have a relatively simple structure, making them relatively straightforward to maintain. However, they are usually less efficient than large turbines. Generally, small turbine technology is used for stand-alone electricity systems and is suitable for rural electrification where a grid connection is not available or required. Wind-diesel hybrid systems can be effective in small or off-grid areas, making use of existing conventional diesel-generating infrastructure while reducing fuel and fuel-transport costs and improving the stability of power supply. Technological innovation is a key factor for future wind power development. Although the technology is relatively mature, further room exists for development. Pilot facilities, for example, are increasingly incorporating energy storage and information technology systems, such as two-way telecommunications between a control centre and remote wind plants, to control power output. To further strengthen wind power development, policy makers should be aware of the latest technological advances.

6. Costs

Despite increases or fluctuations in some cost components, the Levelised Cost of Electricity (LCOE) for wind power has not increased. In some countries where wind conditions are good and where conventional electricity generation costs are high, onshore wind power is cost-competitive with new conventional power plants. The weighted average LCOE for wind power in 2014 was between USD 0.06 per kWh in China and USD 0.12 per kWh in the rest of Asia. The weighted average LCOE in the rest of the world is also within this range, while the best wind projects provide electricity for USD 0.05 per kWh without financial support. Installation and O&M costs are the main elements of the electricity cost for wind power.

For onshore wind, turbine costs dominate, with the rotor blades and tower accounting for nearly half of the total cost of a turbine. After peaking in 2009, turbine prices have declined due to market competition and lower commodity prices. Preliminary turbine price projections for 2014 are USD 676 per kW in China and between USD 931 and USD 1,174 per kW in the United States (IRENA, 2015a). Grid connection costs — including electrical work, electricity lines and connection points — vary depending on the site specifics and on the network or regulatory regime. The main construction cost is for the turbine foundation. Other capital costs include costs for development, engineering and licensing. For onshore wind, the regional weighted average installed cost in 2014 was between USD 1,280 per kW and USD 2,290 per kW. O&M typically accounts for 20%-25% of the LCOE for onshore wind, ranging from USD 0.005 per kWh to USD 0.025 per kWh. Grid connection and construction cost shares are higher for offshore wind than for onshore wind. For a typical offshore wind system in 2014, total installed costs were in the range of USD 2,700-5,070 per kW, and the LCOE was in the range of USD 0.10-0.21 per kWh.

Major factors in reducing the LCOE for wind power are larger turbines and large-scale installation of wind farms. Because larger turbines harness strong wind at higher altitudes, they produce more electricity per unit of installation area, thereby reducing both the number of turbines and the land area needed per unit of output. Large-scale installation of wind farms increases the economies of scale and reduces costs for transport, installation and O&M. Reducing the weight of rotor blades has great potential for reducing turbine costs, as do improving the aerodynamic efficiency and material selection. Carbon fibre is a major candidate for reducing weight and increasing aerodynamic efficiency, but it remains expensive.

The costs of grid connection depend greatly on the site **configuration**. For offshore wind, the potential for reducing the grid connection cost is higher because of the long-distance transmission line needed to connect to the electricity network on land. One option for reducing costs is to use a high-voltage direct current (HVDC) connection. For onshore wind, smart integration of wind power decentralised generation into local and regional grids has the potential to lower system costs substantially, reducing the need for large power networks.

7. Potential and Barriers

Because of the global availability of the resource, wind power has huge potential. An estimated 95 TW or more remains to be developed onshore, and offshore wind has an even larger resource potential, as well as less of an environmental impact. Some countries have introduced financial supports for wind power, such as feed-in tariffs, to secure income and to reduce investor risk. Critical barriers to wind power include long and unpredictable waiting times for permitting and authorization. To reduce such risks, policy makers can introduce appropriate regulatory schemes and set a specific, predictable schedule for the administrative process. Environmental impacts associated with wind development include concerns about noise and visual impact as well as impacts on migratory species. Communication with the public is key to **mitigating** these concerns. Developers need to communicate with stakeholders based on proper environmental impact assessments. Proper siting of wind farms can also **mitigate** visual impacts and impacts on migratory species. Involvement of local communities, particularly through local ownership, is key for high social acceptance.

An important issue for managing power systems that integrate large amounts of wind energy is the variability of the power output. One way to achieve a higher share of wind generation in a grid system is to operate wind turbines or wind farms using integrated transmission systems and power output prediction systems, including weather forecasting. The development of standards and certifications can help to improve the performance of small wind systems, especially in developing countries.

Source:

http://www.iea-etsap.org/Energy-Technologies/Energy-Supply.asp.

 New words and phrases

aerodynamically   *adv.*   空气动力学地

asynchronous   *adj.*   异步的

configuration   *n.*   配置;布局;构造

fluctuate   *v.*   起伏;波动

impose   *v.*   强制执行;强迫接受

kinetic   *adj.*   运动的

mitigate   *v.*   使缓和;使减轻

optimise   *v.*   使最优化

rotational   *adj.*   转动的;回转的;轮流的

rotor   *n.*   转子；水平旋翼

synchronous   *adj.*   同步的；同时的

entrepreneurial spirit   企业家精神

feed-in tariff   （对可再生的）能源补贴

be proportional to   和……成比例的；与……相称的

yaw mechanism   偏航机制；调向机制

 Exercises

I. Discuss the following questions with your partner according to what you read.

1. Why does wind power continue to expand worldwide?

2. What are the three major elements of wind power generation?

3. What concerns do environmental impacts associated with wind development include?

II. Fill in the blanks with the words and phrases in the text.

Efforts are being 1 _____ to improve the economic 2 _____ of wind power facilities. Wind farms are being built to 3 _____ energy production and to 4 _____ capital and operating costs, while remaining within the constraints 5 _____ by the site. Once the site constraints are 6 _____, "micro-siting" is performed to 7 _____ the layout design. For most wind power projects, the economics depend far more on the 8 _____ costs of energy production than on 9 _____ costs. For both onshore and offshore facilities, the 10 _____ parameter for layout design is the maximisation of energy production.

III. Please give the Chinese or English equivalents of the following terms.

1. wind farm

2. electromagnetic induction

3. kinetic energy

4. electric generator

5. fluctuating cost

6.（对可再生的）能源补贴

7. 资金密集型的

8. 商品价格

9. 企业家精神

10. 太瓦

IV. Please translate the following sentences into English.

1. 电动机发明后，工程师们开始利用风能来发电。

2. 2015 年年底，世界上风能储存量前五的国家是中国（148 GW）、美国（74 GW）、德国（45 GW）、印度（25 GW）和西班牙（23 GW）。

3. 风能在运输业、工业和农业领域里已应用数千年了。

4. 全球正在采取措施以提高风能设施的经济效益。

5. 与大多数可再生能源一样，风力发电也需大量资本投入。

V. Please translate the following passage into Chinese.

The costs of grid connection depend greatly on the site configuration. For offshore wind, the potential for reducing the grid connection cost is higher because of the long-distance transmission line needed to connect to the electricity network on land. One option for reducing costs is to use a high-voltage direct current (HVDC) connection. For onshore wind, smart integration of wind power decentralised generation into local and regional grids has the potential to lower system costs substantially, reducing the need for large power networks.

Text B　Socio-ecnomic Benefits of Wind Energy

1. Wind Sector Employment and Local Value Chain

Employment opportunities are a key consideration in planning for low-carbon economic growth. Many governments have **prioritised** renewable energy development, primarily to reduce emissions and meet international climate goals, but also in pursuit of broader socio-economic benefits.

Together, the onshore and offshore wind industries employed 1.16 million people worldwide in 2018. Most wind jobs are found in a small number of countries, although the concentration is less than in the solar PV sector. Asia accounted for nearly half of the global wind employment (620,000 jobs) followed by Europe (28%) and North America (10%). On a country level, China remains the global leader in wind installations, with 44% of global wind employment (510,000 jobs) in 2018. Germany ranked second with 140,800 wind jobs, followed by the US, where wind employment grew 8% to a new peak of 114,000 jobs by the end of 2018.

Arising out of its modeling work that assesses the socio-economic implications of the Remap Case, IRENA estimates that employment in the wind industry would continue to rise, **exceeding** 3.7 million jobs by 2030 and 6 million jobs by 2050. Of the more than 6 million jobs by 2050, 5 million would be in the onshore wind sector and the rest would be offshore jobs (1 million). By project **segment**, of the more than 5 million onshore wind sector jobs by 2050, 1.7 million would be in construction and installation, 2.18 million in manufacturing and 1.17 million in O&M. For offshore wind, out of the total of 1 million jobs by 2050, an estimated 0.45 million would be in construction and installation, 0.39 million in manufacturing and 0.17 million in O&M.

Shifting to a renewable-powered future creates employment opportunities and potentially allows for **retaining** existing expertise from the fossil fuel industry, particularly for renewable technology developments such as offshore wind. For instance, the expertise of workers and technicians in building support structures for offshore oil and gas sites could potentially be utilised to build foundations and substations for offshore wind turbines.

The rising **traction** of the wind energy industry demands a growing array of skills, including technical, business, administrative, economic and legal, among others. Widening the talent pool is thus a **pragmatic** reason for boosting the participation of women in renewable energy, in addition to considerations of greater gender equity and fairness.

2. Wind Projects Create Ample Opportunities for Local Value Creation

Local wind energy industries have the potential to create jobs and develop local manufacturing. Opportunities for domestic value creation can be created at each segment of the value chain, in the form of jobs and income generation for enterprises operating in the country. In the case of domestic industry participation in onshore wind farm development, key aspects such as the labour, materials and equipment requirements of each segment of the value chain need to be analysed. Based on this, opportunities for **leveraging** local labour markets and existing industries can be identified to maximise the domestic value chain. Regional and global market dynamics also strongly influence the decision to pursue domestic industry development (IRENA, 2017).

For the offshore wind industry, domestic manufacturing of the main components of an offshore wind farm, such as the foundation and the substation, as well as parts of the turbine, blades, tower, and monitoring and control system could be considered. Socio-economic gains in terms of local income and jobs can be maximised by leveraging existing economic building on domestic supply chain markets. Sufficient education and training are crucial to build capable local supply chains.

Maximising value creation from the development of a domestic wind industry, for example, requires leveraging capacities in industries such as steel and fibre glass. For example, for a typical 50 MW onshore wind facility, almost 23,000 tonnes of concrete is needed for the foundations, and nearly 6,000 tonnes of steel and iron are needed for the turbines and foundations. The requirements are similar for offshore wind. Manufacturing the main components of a wind turbine requires specialised equipment as well as **welding**, lifting and painting machines that are used in other industries, such as construction and **aeronautics**. The foundations also require the use of specialised equipment including rolling, drilling and welding machinery. Special vessels and cranes are needed to move these big structures. Examining these requirements provides insights on the industrial capabilities to be leveraged.

IRENA's Leveraging Local Capacity report series generates valuable information for policy makers on the occupational and skill structure along the value chain. For example, a total of 144,000 person-days is needed for the development of a 50 MW onshore wind project. The labour requirements are highest in O&M (43% of the total), followed by construction and installation (30%) and manufacturing (17%).

For offshore wind, the majority of the labour requirements (totaling 2.1 million person-days for a 500 MW farm) are found in the manufacturing and **procurement** segment. Existing manufacturing facilities for onshore wind can serve the needs of the offshore sector, as many components are comparable. Significant **synergies** also exist between the offshore oil and gas industry and the offshore wind sector.

3. Clustering with Other Low-carbon Technologies: Hybrid Systems

To overcome the intermittency issue arising from the variable nature of wind energy, and to maintain the reliability and continuous operation of the power system in times of low resource availabilities, a solution would be to combine wind systems with other renewable generation sources such as solar PV, hydro or storage technologies, or with emerging technologies such as hydrogen. In 2012, the world's first hybrid project — combining 100 MW of wind and 40 MW of solar PV generation along with a 36 MW **lithium-ion** energy storage capacity unit — was installed by Build Your Dreams (BYD) and the State Grid Corporation of China in Zhangbei, Hebei Province (JRC, 2014). In 2017, the wind supplier Vestas announced a large-scale hybrid project that combines 43.2 MW of wind and 15 MW of solar generation along with a 2 MW battery storage capacity unit. The global hybrid solar-wind market is expected to grow from more than USD 0.89 billion in 2018 to over USD 1.5 billion by 2025, reflecting a CAGR of nearly 8.5% over the seven-year period (Zion Market Research, 2019). China was the major market for solar-wind hybrid systems in 2018 and is expected to dominate in coming decades. Countries have already deployed a variety of hybrid projects, and a steady rise in such projects can be expected, especially as a complementary solution for solving grid integration issues.

New words

aeronautics	*n.*	航空学,航空工程
cluster	*v.*	使聚集;群聚
	n.	群;簇;丛
exceed	*v.*	超过;胜过
leverage	*v.*	举债经营;借贷收购;利用
lithium-ion	*n.*	锂电子
pragmatic	*adj.*	实际的;实用的
prioritise	*v.*	优先考虑或处理
procurement	*n.*	采购;获得
retain	*v.*	保持;保留;聘请
synergy	*n.*	协同;协同作用;增效
segment	*n.*	部分;片段
traction	*n.*	牵引;拽
weld	*v.*	焊接;使成整体
Compound Annual Growth Rate（CAGR）		年均复合增长率

Exercises

Discuss the following questions with your partners.

1. Make a list of the wind sector employment opportunities nowadays and in the near future.

2. How do the onshore and offshore industries create opportunities for local value creation.

3. What is the meaning of hybrid systems? Say something about the world's first hybrid project.

Keys to Exercises

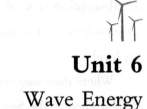

Unit 6
Wave Energy

Lead-in: *The rapid social and economic development has brought about increasingly serious negative effects to our life. Energy shortage, environmental pollution and other problems have seriously hindered people's normal production. How to develop healthily and sustainably has become a problem facing all human beings. As is known to all, the ocean is a huge treasure house of energy, such as tidal energy, wave energy and salt difference energy, among which wave energy has the characteristics of high energy density, wide energy distribution and so on, holding certain advantages in the development of sea energy.*

Text A Wave Energy

In this text, we will learn about wave energy which is full of prospect in the field of clean energy. We will focus on the formation and characteristics of wave energy, technologies to harness wave energy, strategies for wave energy **conversion** and problems with wave power generation. Development and utilization of wave energy in China is also mentioned.

1. Formation Mechanism of Wave Energy

The so-called wave energy refers to the dynamic and potential energy of the ocean

surface waves. The mass and quality of the energy contained in the waves are positively **correlated with** the movement period of the waves, the width of the wave surface and the square of the wave height. At the same time, because wave energy is generated by the **absorption** of wind energy by the ocean, its energy quality, quantity and density are closely related to the wind speed and the effective distance between the wind and sea level.

Where does wave energy come from? Believe it or not, wave power is another form of solar energy. The Sun heats different parts of the globe to different extents, and the resulting temperature differences create the winds that interact with the sea level within the effective acting distance, and part of the water mass produces an effective displacement in the vertical direction, so that the waves have energy. The horizontal motion of the water gives the waves kinetic energy. Solar radiation also creates temperature differences in the water itself, and these drive underwater currents. It may be possible to harness the energy of these currents in the future, but for now, most of the attention of the energy industry has been focused on surface waves.

2. Characteristics of Wave Energy

It can be seen from the formation mechanism of wave energy that the formation and utilization of wave energy depend on external factors that determine the period of wave motion and the height of the maximum wave peak, such as the intensity of air movement at sea level and the influence of density difference in different directions of washing region.

Taking into account a number of factors, wave energy has the following characteristics: (1) the energy density of wave energy is not stable enough; (2) due to the scope of the sea and the impact of severe weather, the development and utilization of wave energy is difficult, and the development cost is relatively high; (3) wave energy is greatly affected by external factors, so it is difficult to achieve effective stability of power generation frequency and voltage, which requires higher energy conversion device.

3. Technologies to Harness Wave Energy

The bulk of electricity that powers the industrial world comes from induction generators. The first one came online in 1896 and was powered by the falling cascade of water that is Niagara Falls. Most modern induction generators are steam-driven, though, and the fuels of choice to heat the water have long been coal, petroleum and natural gas — so-called fossil fuels.

As of 2011, fossil fuels supplied 82 percent of the world's electricity, but evidence continues to mount of the **devastating** effects the byproducts of combustion have on the environment. As of October 2018, scientists were warning that global warming, to which fossil fuel combustion is a prime contributor, was quickly approaching an **irreversible** tipping point. The result of such warnings is a shift away from fossil fuels and toward renewable energy sources, such as photovoltaic panels, **geothermal energy** and wind turbines.

Wave power is one of the options on the table. The oceans represent a vast reservoir of untapped energy. According to the Electric Power Research Institute, the potential wave energy around the coastal United States, including Alaska, is around 2,640 terawatt-hours/year. That's enough energy to power to 2.5 million households for an entire year. Another way to look at it is that a single wave has enough energy to power an electric car for hundreds of miles.

Four main technologies exist to harness wave energy, some work near the shore, some offshore and some in the deep sea. Wave energy converters (WECs) are designed to remain on the surface of the water, but they differ in the **orientations** of the collectors to the movement of the waves and in the methods used to generate electricity. Some devices generate the electricity on the spot and transmit it via undersea cables to shore, while others pass the mechanical energy of the wave along to land before turning it into electrical energy. Which of these drastically divergent concepts might emerge as a winner is far from clear.

The four types of wave electricity generators currently existing include **point absorbers**, **terminators**, **overtopping devices** and **attenuators**.

3.1 Point absorbers resemble buoys

A point absorber is primarily a deep-sea device. It remains anchored in place and **bobs** up and down on the passing waves. It consists of a central **cylinder** that floats freely inside a housing, and as the wave passes, the cylinder and housing move relative to each other. The motion drives an electromagnetic induction device or a **hydraulic**

piston, which creates the energy necessary to drive a turbine. Because these devices absorb energy, they may affect the characteristics of the waves that reach shore. This is one reason why most of them are usually used in locations offshore.

Differential pressure devices are submerged point absorbers placed near shore and attached to the seabed by a single point, and they use the pressure difference generated between the crest and the trough of the wave to create an up and down motion of the device.

An **oscillating water column** (OWC, see Fig. 6-1) is a particular type of point absorber. It also looks like a buoy, consisting of a submerged air chamber connected to the atmosphere by means of a circular duct inside which a bidirectional flow turbine is installed. The flow rate of air through the turbine is generated by the successive incident sea water waves that **compress** and depressurize air in the air chamber by means of the periodic motion of the oscillating free surface.

Fig. 6-1 Pictures for OWC

http://www. 360doc. com/content/14/0107/19/12109864_343394176. shtml.

3.2 Terminators generate wave electricity from compressed air

Terminators can be located on shore or near the shoreline. They are basically long tubes, and when deployed offshore, they capture water through subsurface port openings. The tubes are anchored to extend in the direction of wave motion, and the rise and fall of the ocean surface pushes a column of captured air through a small opening to drive a turbine. When they are located onshore, the waves crashing onto the beach drive the process, so the openings are located in the ends of the tubes. Each terminator can generate power in a range from 500 kilowatts to 2 megawatts, depending on wave conditions. That's enough power for an entire neighborhood.

3.3 Attenuators are multi-segmented wave energy converters

Like terminators, attenuators are long tubes that are deployed **perpendicular** to the wave movement. They are anchored at one end and constructed in **segments** that move relative to one another as the wave passes. The movement drives a hydraulic piston or some other mechanical device located at each segment, and the energy drives a turbine, which in turn produces electricity.

3.4   Overtopping devices are like mini hydroelectric dams

Overtopping devices are long and extend perpendicular to the direction of wave motion. They form a barrier, much like a seawall or dam, that collects water. The water level rises with each passing wave, and as it falls again, it drives turbines that generate electricity. The overall action is roughly the same as the one employed in hydroelectric dams. The turbines and transmission equipment are often housed in offshore platforms. Overtopping devices can also be constructed onshore to capture the energy of waves that crash onto the beach.

4. Strategies for Wave Energy Conversion

In a hydroelectric dam, the energy of falling water directly **spins** the turbines that generate **AC electricity**. This principle is used almost **unaltered** in some forms of wave generation, but in others, the energy of the rising and falling water has to pass through another medium before it can do the work of spinning the turbine. This medium is often air. The air is sealed in a **chamber**, and the motion of the waves compresses it. The compressed air is then forced through a small **aperture**, creating a jet of air that can do the necessary work. In some technologies, the energy of the waves is transferred to mechanical energy by hydraulic pistons. The pistons in turn drive the turbines that generate electricity.

Wave power is still largely in the experimental phase, and hundreds of different designs have been patented, although only a fraction of these have actually been developed. The process to refine the technologies is ongoing. The European Marine Energy Centre in Scotland's Orkney Islands allows companies to connect their devices to existing infrastructure and cabling to test their electricity-generating capabilities and identify problems. Australian company Carnegie Wave plans to commission a "commercial scale" installation near Perth, using a fully submerged device that uses wave power to pump water to shore for conversion to electricity. And there are signs that big-company buy-in is starting, as evidenced by Lockheed Martin's Australia project,

新能源电力英语
English for Renewable Electricity Sources

which will use a buoy technology that generates electricity from the rising and falling of waves.

5. Problems with Wave Power Generation

Despite the obvious promise of wave power, development lags far behind that of solar and wind power. Large-scale commercial installations are still a thing of the future. Some energy experts **liken** the state of wave electricity to that of solar and wind electricity 30 years ago. At that time, engineers had not settled on the optimal design for wind turbines, but decades of ensuing research had resulted in highly sophisticated turbine designs. With wave power, some research occurred after the Arab oil embargo of the 1970s, but since then government and commercial research and development into wave power has paled compared to wind and solar energy. Part of the reason for this is **inherent** in the nature of ocean waves which are irregular and unpredictable. The height of the waves and their period, which is the space between them, can vary from day to day or even hour to hour.

From a technical point of view, operating in the ocean is far more difficult than on land; building offshore wind installations, for example, tends to be significantly more expensive than constructing wind farms onshore. Saltwater is a hostile environment for devices, and the waves themselves offer a challenge for energy harvesting as they not only roll past a device but also bob up and down or converge from all sides in confused seas. This provides enticing opportunities for energy capture, but a challenge for optimum design.

Another problem is power transmission. Wave power can't serve any purpose until it is transmitted to the shore. Most WECs incorporate transformers to step up the voltage for more efficient transmission along underwater power lines. These power lines typically rest on the sea bed, and installing them adds significantly to the cost of a wave power generation station, especially when the station is located far from shore. Moreover, there is a certain amount of power loss associated with any transfer of electric energy.

6. Development and Utilization of Wave Energy in China

The long coastline and vast sea area determine that China has abundant wave energy and great development potential. The development and utilization of wave energy in China began in the 1970s. In 1975, a 1000-W wave energy generating device was made and put into the test, which was improved and upgraded. **In terms of** the construction of wave power stations, Guangzhou Energy Institute built a 3,000-W multi-oscillating water-column wave power station in 1989, and successfully tested the power generation in 1996 and upgraded it to a 20-kW wave power station. Shanwei city, Guangdong Province, built the world's first independently stable wave power station in 2005. In recent years, Shandong University, Ocean University of China, National Marine Technology Center and other institutions have made technological breakthroughs.

To sum up, current wave energy development situation, whether from the technical level or from the aspect of scale, has yet to reach mass concentration degree of power, but the wave energy power generation industry has been developing steadily over the years. Solving related technical problems, the development of wave energy industry will make a significant contribution to the world's energy industry.

Source：

1. https://sciencing.com/how-is-hydropower-gathered-or-created-13660670.html.

2. https://sciencing.com/waterfall-generate-power-4683787.html.

3. https://sciencing.com/wave-energy-used-generate-electricity-6499297.html.

4. https://sciencing.com/facts-5778942-hydropower-non-renewable-renewable-resource-.html.

5. http://www.haiyangkaifayuguanli.com/ch/reader/view_abstract.aspx?file_no=20180413&flag=1.

6. https://e360.yale.edu/features/why_wave_power_has_lagged_far_behind_as_energy_source.

7. https://link.springer.com/article/10.1007/s11804-018-00060-8.

New words and phrases

absorption　　*n.*　　吸收;全神贯注,专心致志

aperture　　*n.*　　孔,穴;(望远镜等的)光圈;缝隙

attenuator　　*n.*　　衰减器;弱化子

bob　　*v.*　　(使)上下跳动,振动;(沿某个方向)快速移动

chamber　　*n.*　　(身体或器官内的)室,膛;房间;会所

compress　　*v.*　　压缩;压紧;精简

conversion   *n.*   转换;变换;兑换;改变信仰

cylinder   *n.*   圆筒;汽缸

devastating   *adj.*   毁灭性的;全然的

hydraulic   *adj.*   液压的;水力的;水力学的

inherent   *adj.*   固有的;内在的;与生俱来的,遗传的

irreversible   *adj.*   不可逆的;不能取消的;不能翻转的

liken   *v.*   比拟;把……比作

orientation   *n.*   方向;定向;适应;情况介绍

oscillate   *v.*   振荡;摆动;犹豫

perpendicular   *adj.*   垂直的;陡峭的;直立的

piston   *n.*   活塞

segment   *n.*   段;部分

spin   *n.*   旋转;疾驰

　　 v.   (使)旋转;纺纱

terminator   *n.*   终结者;明暗界限

unaltered   *adj.*   不变的;未被改变的;照旧的

AC electricity   交流电(AC 即 alternating current)

be correlated with   与……有关

geothermal energy   地热能

in terms of   依据;按照;在……方面;以……措词

oscillating water column (OWC)   震荡水柱

overtopping device   漫顶装置

point absorber   点式波能吸收器

take... into account   考虑;重视;体谅

Exercises

I. Read the text and discuss over the following questions with your partners.

1. How is wave energy formed? Please describe it according to what you learn in Text A.

2. Is it easy or difficult to put wave energy into practice? Share your reasons.

3. What do you think of the four technologies introduced in Text A? Choose one of them, surf the Internet for more information and make a presentation in your class.

Ⅱ. Fill in the blanks with the words and phrases in the text.

In a hydroelectric dam, the energy of falling water directly 1 _____ the turbines that generate AC 2 _____. This principle is used almost 3 _____ in some forms of wave 4 _____, but in others, the energy of the rising and falling water has to pass through another 5 _____ before it can do the work of spinning the turbines. This medium is often air. The air is sealed in a 6 _____, and the motion of the waves 7 _____ it. The compressed air is then forced through a small aperture, creating a jet of air that can do the necessary work. In some technologies, the energy of the waves is 8 _____ to mechanical energy by hydraulic 9 _____. The pistons in turn drive the 10 _____ that generate electricity.

Ⅲ. Please give the English or Chinese equivalents of the following terms.

1. wave peak

2. wave energy converter

3. point absorber

4. oscillating water column

5. air chamber

6. 漫顶装置

7. 太阳辐射

8. 形成机制

9. 地热能

10. 电力传输

Ⅳ. Please translate the following sentences into English.

1. 由于波浪能是由海洋吸收风能而产生的,其质量、数量及密度跟风速、风与海平面的有效距离密切相关。

2. 太阳把地球上不同的地带加热到不同的程度,温差产生的风在有效距离内与海平面相互作用,部分水体在垂直方向产生有效的位移,由此产生了波浪能。

3. 由波浪能的形成机制可以看出,波浪能的形成和利用取决于决定波浪的周期波动和最大波峰高度的外部因素,如海平面空气运动的强度和不同方向的波浪所产生的密度差。

4. 截至 2018 年 10 月,科学家们警告称,全球变暖正迅速接近一个不可逆转的临界点。

5. 波浪能只有输入到岸上才能发挥作用。

V. Please translate the following passage into Chinese.

Despite the obvious promise of wave power, development lags far behind that of solar and wind power. Large-scale commercial installations are still a thing of the future. Some energy experts liken the state of wave electricity to that of solar and wind electricity 30 years ago. Part of the reason for this is inherent in the nature of ocean waves. They are irregular and unpredictable. The height of the waves and their period, which is the space between them, can vary from day to day or even hour to hour.

Another problem is power transmission. Wave power can't serve any purpose until it is transmitted to the shore. Most WECs incorporate transformers to step up the voltage for more efficient transmission along underwater power lines. These power lines typically rest on the sea bed, and installing them adds significantly to the cost of a wave power generation station, especially when the station is located far from shore. Moreover, there is a certain amount of power loss associated with any transfer of electric energy.

Text B   Ocean Wave Energy Research and Development in China

China is the largest energy-consuming country in the world at present, **accounting for** approximately one-fifth of the world's total energy consumption. By 2030, China's energy consumption is expected to increase by 60%. Thus, China's energy policy will affect the energy direction of the entire world. Fortunately, China has vast resources, and there is great potential for marine renewable energy (MRE), including energy in the Bohai, Yellow, East China and South China seas along the country's coastline from north to south.

As a type of promising renewable energy, marine renewable energy, which is promoted by the Chinese government and plays an important role in supporting the adjustment of the national industrial structure, is expected to contribute to a global optimization of the energy mix, **fostering** new industries and exploring reformation of its energy mix. During the past few decades, specific strategies, plans and national funding for MRE have been implemented by the Chinese government, and research, development, and deployment (RD&D) related to MRE has been rapidly occurring. Among the most promising ocean energy resources, technologies for wave energy have also achieved great progress.

The **initiation** of wave energy converter (WEC) research and engineering application in China dates back to the late 1970s. At first, only a few research institutes carried out studies in this field. However, the current situation has **substantially** changed in mainly three aspects. First, at the government level, the central and local governments have issued a number of policy documents to stimulate the progress of MRE. Meanwhile, a series of national incentives and funds have been supporting acceleration of the process of demonstration and commercialization. Second, at the academic level, many institutes and universities are carrying out a large number of research projects (from small-scale indoor wave basin experiments to **prototype** sea trials) related to MRE and have published many articles and **patents**. Finally, at the commercial level, some companies are developing MRE technologies independently or jointly with universities for commercial purposes. Some engineering prototypes have

121

新能源电力英语
English for Renewable Electricity Sources

been constructed, and most have been tested in the sea. Additionally, several sea test sites are being built in China for public sharing.

1. Wave Energy Resource in China

The coastal seas of China are in the East Asian monsoon region. Waves mainly **originate from** the north during winter because of the cold and dry **prevailing** northern winds from high-latitude areas. However, during summer, the seas are dominated by southern winds and waves affected by the warm and humid currents from the Pacific Ocean and South China Sea. Generally, the wave period is relatively small, and the wave length is short. With the change in seasons, winter waves are larger than summer waves.

Compared with other countries, the wave power is relatively low along China's coastline. The average wave power density is less than 10 kW per meter of wave front, which is much less than that between 40° and 60° in latitude, such as in Australia, the United States, South Africa, and Western European countries. Thus, different measures should be taken to design WECs that are suitable for China's low wave power density. In other words, the WEC style and scale size in western countries may not be **applicable** to China, and some small-scale and high-conversion-efficiency WEC devices should be the focus.

In China, the **spatial** distribution of ocean wave power is **uneven**. The annual mean wave power density increases from north to south and from the nearshore to the offshore along China's coastline, and concretely, it is less than 2 kW/m in most areas of the Bohai Sea. For the northern and nearshore areas of the Yellow Sea, the average value is 1-2 kW/m, while it is 2-3 kW/m offshore for the southern part. For the East China Sea, the wave power density is greater than that in the Bohai and Yellow seas, where the Zhoushan **archipelago** in northern Zhejiang Province is generally greater than 2 kW/m, and for the southern part, such as on Dachen Island, it is greater than 3 kW/m. For the northern part of the South China Sea along the coastline of Guangdong Province and Hainan Island, this value is generally between 3 and 5 kW/m.

According to the distribution of the annual mean wave power density of Chinese coastlines, four grade divisions can be **designated**, which are high, moderate, exploitable and poor, respectively. Southern Fujian, northern Guangdong, southwest

Hainan, and Taiwan have the most abundant coastal areas, where the wave power is relatively high, while seas in Liaoning, Hebei, Tianjin, Shandong, and Jiangsu have poor wave power resources.

As per the data of the 908 Special Fund for China's offshore marine renewable program, started in 2004, the wave power distribution of different regions 20 km from the coastline of China apart from Taiwan is also of great difference. The total theoretical wave power reaches up to 16 GW, and the technically exploitable wave power is nearly 14.71 GW. Guangdong holds the richest exploitable wave power, followed by Hainan and Fujian, while Tianjin possesses the least.

2. Strategies and Plans

With construction of the "Twenty-first Century **Maritime** Silk Road, " "Building China into a Sea Power Nation, " "One Belt, One Road" initiatives and other national strategies, the State Council, National Energy Administration (NEA) and State Oceanic Administration (SOA) of China have implemented a series of policies and plans to exploit, utilize, and protect ocean resources. Thus, a positive policy environment to promote the development of MRE has been established.

In March 2016, the Chinese government released the "Outline of the 13th Five-Year Plan for Economic and Social Development", which emphasized the construction of a modern energy system, development of blue economic space, protection of the marine environment, and breakthrough of technology and techniques in the MRE field. In December 2016, SOA issued the "13th Five-Year Plan (2016-2020) for Marine Renewable Energy," aiming to promote the industrialization of marine energy technology and accelerate the process of commercialization. **Notably**, by 2020, more than five island projects for multi-energy complementary power supply stations based on MRE are scheduled to be completed, with a 50 MW total target rated capacity of MRE being built, in which a 500-kW wave energy demonstration power station is included. **Simultaneously**, a group of efficient, stable, and reliable technical products will be created by the end of the 13th Five-Year Plan. The "Action Plan for Energy Technology Revolution and Innovation (2016-2030), " **formulated** jointly by the National Development and Reform Commission (NDRC) and NEA in May 2016, is expected to result in the development of high-efficiency WECs, construction of MRE

demonstration test sites, and establishment of a successive supply chain for MRE by 2030.

In addition to the aforementioned central national policies, some coastal provinces and cities have also promulgated some local regulations and management measures to stimulate the development of local MRE resources, depending on local conditions. There are also some other local incentive documents associated with MRE that are not listed **herein**. All of these local policies in line with the central government strategies have greatly and rapidly pushed China's MRE technology and techniques forward.

3. Technical Research and the Development of WEC

3.1 Organizations engaged in WEC in China

The level of knowledge and expertise **regarding** WECs in China is immature compared to that in western countries, but the interest is extremely high. Over the past ten years, many organizations have been involved in the R&D of WEC under national policy incentives and financial support in China. In addition, traditional organizations such as research institutions and universities and increasingly more state owned and private enterprises have been actively participating in this field during recent years, particularly since the establishment of the Special Funding Plan for Marine Renewable Energy (SFPMRE). According to statistics, there are currently more than 30 organizations involved in the research and exploitation of WECs, though this is far from sufficient compared to some western countries.

3.2 WEC types

Currently, the utilization of WECs in China is in the early stage from technical research to commercial application. The core technology has not yet made a fundamental breakthrough, and problems such as low device conversion efficiency and poor stability and reliability have remained bottlenecks restricting the development of WECs. However, during the "12th Five-Year Plan," China independently designed and **deployed** more than 40 devices for wave energy conversion, most of which were verified in the laboratory and tested in actual seas. Moreover, more than 200 scientific research articles have been published, and more than 150 patents have been granted related to wave energy conversion in China. Nearly all the WEC types, including oscillating water column (OWC), oscillating bodies (OB), and overtopping (OT), as well as some other types such as flap-type, **magneto hydrodynamic** generation (MHD), and multi-energy complementary power generation, have been studied in China with the following as representatives.

The study of OWC in China dates back to the 1980s, when it was regarded as among the most promising types in the WEC community. In China, before 2002, WEC research was mainly focused on the OWC type, and a variety of devices of this type, such as navigation buoys and navigation ships, were successfully deployed in real seas. Of these devices, navigation buoys used in the Yangtze Estuary Waterway have been commercialized and exported to Japan and other countries.

Generally, according to the power generation principle of WEC devices, most devices currently under study belong to the category of oscillating body (OB) WECs. This type of device is characterized by the relative motion between a single floating body and the seafloor or a **plurality** of floating bodies. The representatives of such a WEC in China are undoubtedly the series of devices "Duck" and "Sharp Eagle", two devices named for the shapes of their primary-phase energy conversion buoys that resemble the head of a duck and an eagle, which have been continuously operated by the Guangzhou Institute of Energy Conversion (GIEC). In addition, the "Ne Zha I" and "Ne Zha II" were also designed and deployed by the GIEC. A hydraulic system was adopted as the energy transmission medium in all of the aforementioned WEC systems.

Two main categories of MRE multi-energy synergy systems at the moment are **co-located** systems and **hybrid** systems according to whether the same platform is shared. A co-located system combines two or more MRE resources in the same ocean space but with independent platforms or structures. However, a hybrid system is a multipurpose platform combining different MRE resources, for example, offshore wind turbines and WECs, together in the same structure or platform. Both synergy systems share the relevant **infrastructures** and consequently reduce the overall cost. To date, the hybrid system has been given additional research, and the wind-wave synergy is the most advanced and promising type, mainly due to the high maturity level of the offshore wind turbine and the abundant wave energy resources.

3.3　Levelized cost of MRE

The **levelized cost** of energy (LCOE) is defined as the unit cost of electrical outputs over the lifetime of a specific energy source. From the very beginning, studies on MRE have focused on improving the efficiency and reliability of the device itself. However, little attention has been paid to the cost, which is also a key factor affecting commercial application. In recent years, studies on the levelized cost of MRE have been given more attention and discussion, but mainly in western countries. At present, different models are used to calculate the levelized cost of MRE, and the discounting method is mostly used. The results show that the levelized costs are approximately €165/MWh, €190/MWh and €225/MWh on average for offshore wind, tidal stream and wave energy, respectively, at 2015 prices. It is found that the wave energy has a higher cost than the offshore wind and tidal stream energy. However, the levelized cost of offshore wind and tidal stream energy are still higher than traditional nonrenewable energy resources, which has been an apparent obstacle of the further development of the technologies.

LCOE is a widely recognized approach worldwide that is adopted to evaluate the economic performance of an energy resource. However, few studies are available on the current state of levelized cost of MRE projects in China. Thus, to date, no **relevant** experimental data and policy standards have been developed to assess the cost performance of MRE projects in China, despite **extensive** field tests conducted in the real sea. Of course, as is known at present, all R&D efforts are supported by government funding and policies. Therefore, future studies should focus on the levelized cost of MRE. Additionally, in China, the current levelized cost of both onshore wind turbine and solar PV is approximately $90/MWh on average. It is projected that grid parity can be achieved by 2020, and these devices will be significantly less costly than traditional energy in the 2050s.

4. Financial Support

Sustained financial support is a significant factor in the technical progress of WECs. Currently, ocean wave energy development in China is in a **critical** period of technological breakthroughs. A series of financial support programs have been implemented by the national central departments as well as some local governments in recent years to promote the overall development of fundamental research, engineering demonstrations, and standard system construction in the MRE field. The following are some representative fund programs.

4.1　Special Fund Plan for Marine Renewable Energy

In May 2010, with the support of the Ministry of Finance, the SOA established the Special Fund Plan for Marine Renewable Energy (SFPMRE). By the end of April 2017, the total amount of financial support reached approximately $148.1 million at 2017 prices, and altogether, 103 projects were supported by the SFPMRE to promote the RD&D of MRE, of which 62 projects passed the official acceptance. This fund mainly focuses on the following five key research fields:

Demonstration of isolated MRE systems;

Demonstration of grid-connected MRE systems;

Industrialization demonstration of MRE systems;

Theoretical and experimental research on MRE systems;

Standard and support service construction of MRE systems.

Among these, the fourth field had the most projects, accounting for as much as 50%, while the fifth field had the largest amount of funding, with a proportion of 41%.

4.2　National Natural Science Foundation of China

During recent years, the National Natural Science Foundation of China (NNSFC) has continuously increased its support for the study of basic scientific issues in the MRE field and has effectively promoted its progress in China. According to data published by the official website of the NNSFC, a lot of research projects related to MRE have been supported in the past few years, with a total funding of over $900 thousand, among which some are general projects, some are youth science foundation projects, and others are international (regional) cooperation and exchange projects and an emergency management project.

4.3　National High-tech R & D Program of China (863 Program) and the National Key Technology Support Program

Both the National High-tech R&D Program of China (863 Program) and the National Key Technology Support Program are by the Ministry of Science and Technology of China. Key technologies of the island multi-energy complementary power

generation system, as well as the wave power generation technology, were **successively** supported by the "863 Program" and the National Key Technology Support Program during the last few decades. Among the main directives of renewable energy technology, the "863 Program" arranged for nearly $1.5 million of research funding during the "11th Five-Year Plan" period. At the same time, $4.8 million were funded by the National Key Technology Support Program for the research and demonstration of key technologies for marine energy development and utilization. The key technologies of a 100-kW floating wave power station and integrated test technologies of a marine energy generation system were the main focus.

5. International Cooperation and Activities

Since 2006, the Chinese government has cooperated with the Global Environment Facility and the World Bank to develop the "China Renewable Energy Scale-up Programme" (CRESP) to support China's efforts to develop and expand renewable energy. During the first **phase** of CRESP (2006-2011), policy implementation, technology improvement, and provincial-level project demonstration were mainly conducted with a grant of $40.22 million. A total of $27.28 million was financed in CRESP II (2013-2019), and large-scale renewable energy generation and consumption, innovation mechanisms, and engineering demonstrations are making good progress.

The International Energy Agency (IEA) and the International Electrotechnical Commission (IEC) have established branches of MRE to accelerate its advancement globally. The relevant departments in China have actively communicated with these organizations, joined the relevant international organizations, and conducted a wide range of cooperative activities and exchanges that have effectively promoted the progress of China's marine technology and standards.

For MRE test site construction and evaluation of WECs, China has actively cooperated with many western countries with advanced marine technology, particularly the UK, Spain, Canada, and Singapore.

In short, China has made great progress in terms of MRE, but some challenges remain need more attention, particularly for WECs. It is globally acknowledged that though a long-lasting and tortuous way is unavoidable, essential and meaningful measures and supports are bound to enhance a bright future on the path of wave energy commercialization.

Source:

1. https://www.sciencedirect.com/journal/renewable-and-sustainable-energy-reviews.

2. https://www.sciencedirect.com/science/article/abs/pii/S1364032119304794? via%3Dihub.

 New words and phrases

applicable　*adj.*　可适用的;可应用的;合适的

archipelago　*n.*　群岛,列岛;多岛的海区

co-locate　*v.*　位于同一位置;同地协作

critical　*adj.*　危险的;决定性的;评论的

deploy　*v.*　配置;展开;部署

designate　*v.*　指定;把……定名为

extensive　*adj.*　广泛的;大量的;广阔的

formulate　*v.*　规划;明确地表达

foster　*v.*　促进;培养

parity　*n.*　平价;同等;相等

herein　*adv.*　本文中,本书中;于此

hybrid　*adj.*　混合的;杂种的

infrastructure　*n.*　基础设施;公共建设;下部构造

initiation　*n.*　启蒙,传授;开始;入会

maritime　*adj.*　海运的,航海的,海事的;近海的,沿海的

notably　*adv.*　显著地;尤其

patent　*n.*　专利权;专利证;专利品

phase　*n.*　月相;时期,阶段

plurality　*n.*　多数;复数;兼职;胜出票数

prevailing　*adj.*　流行的;一般的,最普通的;占优势的

prototype　*n.*　原型;标准,模范

regarding　*prep.*　关于,至于;就……而论

relevant　*adj.*　相关的;切题的

simultaneously　*adv.*　同时地

spatial　*adj.*　空间的;受空间条件限制的

substantially　*adv.*　实质上;大体上;充分地

successively　*adv.*　相继地;依次地

sustain　*v.*　维持;支撑

uneven　*adj.*　不均匀的;不平坦的

account for　对……做出解释;说明……的原因;(数量、比例上)占

levelized cost　平准化成本

magneto hydrodynamic　磁流体动力

originate from　发源于

Exercises

Please discuss over the following questions with your teammates after reading the text. You can support your view with more information from online or other channels.

1. Please make a brief introduction about the characteristics of the wave energy of China.

2. What do you think of the present policies implemented by Chinese governments of all levels to enhance the development of wave energy in China?

Keys to Exercises

3. If you want to contribute to the development of the WECs, what type will you choose from those presented in the text? Surf the Internet to make a presentation and share with your classmates.

Unit 7

Tidal Energy

Lead-in: *Tides are the result of the interaction of the gravity of the Sun, Earth, and Moon. The rise and fall of the tides — in some cases more than 12 m — creates potential energy. The flow due to flood and ebb currents creates kinetic energy. Both forms of energy can be harvested by tidal energy technologies as renewable energy. Tidal energy technologies are not new: examples were already reported in Roman times and ruins of installations — tidal mills — are found in Europe from around the year 700. Since the 1960s, only five projects had been developed commercially in the period up to 2012. However, new technologies have advanced considerably over the past few years and there are a number of ongoing full-scale demonstration projects. Talk with your teammates about the usage of tide and report to your class. The picture below may give you some hints.*

Text A   Tidal Power

In this text, we will learn about tidal energy used as electricity sources for the benefit of human beings. A range of technologies, including tidal power, will be needed for clean energy transitions around the world. The text will focus on what tidal power is, how it is generated, the cost indication of tidal energy, the advantages and

131

disadvantages of it, the differences between tidal energy and wave energy and the potential and future of tidal energy.

1. What Is Tidal Energy?

Tidal energy is a form of hydropower that converts the energy obtained from tides into useful forms of power, which is produced by the **surge** of ocean waters during the rise and fall of tides. Tidal energy is a renewable source of energy.

During the 20th century, engineers developed ways to use tidal movement to generate electricity in areas where there is a significant tidal range — the difference in area between high tide and low tide. All methods use special generators to convert tidal energy into electricity.

Tidal energy production is still in its **infancy**. The amount of power produced so far has been small. There are very few commercial-sized tidal power plants operating in the world. The first was located in La Rance, France. The largest facility is the Sihwa Lake Tidal Power Station in South Korea. The United States has no tidal plants and only a few sites where tidal energy could be produced at a reasonable price. China, France, England, Canada and Russia have much more potential to use this type of energy.

In the United States, there are legal concerns about underwater land ownership and environmental impact. Investors are not enthusiastic about tidal energy because there is not a strong guarantee that it will make money or benefit consumers. Engineers are working to improve the technology of tidal energy generators to increase the amount of energy they produce, to decrease their impact on the environment, and to find a way to earn a profit for energy companies.

2. How Is Tidal Energy Generated?

There are currently three different ways to get tidal energy: tidal streams, barrages, and tidal **lagoons**.

2.1 Tidal stream

For most tidal energy generators, turbines are placed in tidal streams. A tidal stream is a fast-flowing body of water created by tides. A turbine is a machine that takes energy from a flow of fluid. That fluid can be air (wind) or liquid (water). Because water is much more dense than air, tidal energy is more powerful than wind energy.

Unlike wind, tides are predictable and stable. Where tidal generators are used, they produce a steady, reliable stream of electricity.

Placing turbines in tidal streams is complex, because the machines are large and **disrupt** the tide they are trying to harness. The environmental impact could be severe, depending on the size of the turbine and the site of the tidal stream. Turbines are most effective in shallow water. This produces more energy and allows ships to navigate around the turbines. A tidal generator's turbine blades also turn slowly, which helps **marine** life avoid getting caught in the system.

The world's first tidal power station was constructed in 2007 at Strangford Lough in Northern Ireland. The turbines are placed in a narrow strait between the Strangford Lough inlet and the Irish Sea. The tide can move at 4 meters per second across the strait.

2.2   Tidal barrage

Another type of tidal energy generator uses a large dam called a barrage. With a barrage, water can spill over the top or through turbines in the dam because the dam is low. Barrages can be constructed across tidal rivers, bays, and estuaries.

Turbines inside the barrage harness the power of tides the same way a river dam harnesses the power of a river. The barrage gates are open as the tide rises. At high tide, the barrage gates close, creating a pool, or tidal lagoon. The water is then released through the barrage's turbines, creating energy at a rate that can be controlled by engineers.

The environmental impact of a barrage system can be quite significant. The land in the tidal range is completely disrupted. The change in water level in the tidal lagoon might harm plant and animal life. The **salinity** inside the tidal lagoon lowers, which changes the organisms that are able to live there. As with dams across rivers, fish are blocked into or out of the tidal lagoon. Turbines move quickly in barrages, and marine animals can be caught in the blades. With their food source limited, birds might find different places to migrate.

A barrage is a much more expensive tidal energy generator than a single turbine. Although there are no fuel costs, barrages involve more construction and more machines. Unlike single turbines, barrages also require constant supervision to adjust power output.

The tidal power plant at the Rance River **estuary** in Brittany, France, uses a barrage. It was built in 1966 and is still functioning. The plant uses two sources of energy: tidal energy from the English Channel and river current energy from the Rance River. The barrage has led to an increased level of **silt** in the habitat. Native **aquatic** plants **suffocate** in silt, and a flatfish called plaice is now extinct in the area. Other organisms, such as cuttlefish, a relative of squids, now thrive in the Rance **estuary**. Cuttlefish prefer cloudy, silty ecosystems.

2.3 Tidal lagoon

The final type of tidal energy generator involves the construction of tidal lagoons. A tidal lagoon is a body of ocean water that is partly enclosed by a natural or man-made barrier. Tidal lagoons might also be estuaries and have freshwater emptying into them.

A tidal energy generator using tidal lagoons would function much like a barrage. Unlike barrages, however, tidal lagoons can be constructed along the natural coastline. A tidal lagoon power plant could also generate continuous power. The turbines work as the lagoon is filling and emptying.

The environmental impact of tidal lagoons is **minimal**. The lagoons can be constructed with natural materials like rock. They would appear as a low breakwater (sea wall) at low tide, and be **submerged** at high tide. Animals could swim around the structure, and smaller organisms could swim inside it. Large predators like sharks would not be able to penetrate the lagoon, so smaller fish would probably thrive. Birds would likely flock to the area.

But the energy output from generators using tidal lagoons is likely to be low. There are no functioning examples yet. China is constructing a tidal lagoon power plant at the Yalu River, near its border with North Korea. A private company is also planning a small tidal lagoon power plant in Swansea Bay, Wales.

3. The Cost Indications of Tidal Energy

Tidal range power generation is dominated by two large plants in operation, the La Rance Barrage in France and the Sihwa Dam in South Korea. The construction costs for La Rance Barrage were around USD 340/kW (2012 value, commissioned in 1966), whilst the Sihwa dam was constructed for USD 117/kW in 2011. The latter used an existing dam for the construction of the power generation technology. The construction cost estimates for proposed tidal barrages range between USD 150/kW in Asia to around USD 800/kW in the UK, but are very site specific. Electricity production costs for La Rance Barrage and Sihwa Dam are EUR 0. 04/kWh and EUR 0. 02/kWh, however, these costs are very site specific. Tidal range technologies can be used for coastal projection or water management, which would reduce the upfront costs. On the other hand, additional operational costs may occur due to the control, monitoring and management of the ecological status within the **impoundment**.

Tidal current technologies are still in the demonstration stage, so cost estimates are projected to decrease with deployment. Estimates from across a number of European studies for 2020 for current tidal technologies are between EUR 0. 17/kWh and EUR 0. 23/kWh, although current demonstration projects suggest the levelised cost of energy (LCOE) to be in the range of EUR 0. 25-0. 47/kWh. It is important to note that costs should not be considered as a single performance indicator for tidal energy. For example, the costs for both tidal range and tidal stream technologies can fall by up to 40% in cases where they are combined and integrated in the design and construction of existing or new infrastructure.

4. The Advantages and Disadvantages of Tidal Energy

4.1　Advantages of tidal energy

It is renewable. Tidal energy's source is a result of the effects of the Sun and Moon's gravitational fields, combined with our planet's rotation around its **axis**, which results in low and high tides. With this in mind, the power source of tidal energy is potentially renewable, whether we are talking about tidal barrages, stream generators, or the more recent technology, dynamic tidal power (DTP). Compared to nuclear reserves and fossil fuels, the Sun and Moon's gravitational fields as well as the Earth's rotation around its axis will not cease to exist any time soon.

It is green. Aside from being renewable, tidal energy is also an environmentally friendly energy source because it does not take up a lot of space and does not emit any greenhouse gases. However, there are already some examples of tidal power plants and their effects on the environment. Important studies and assessments are being conducted on these things.

It is predictable. Sea currents are highly predictable, developing with well-known cycles. This makes it easier to construct tidal energy systems with the correct dimensions because the kind of power the equipment will be exposed to is already known. This is why both the equipment's installed capacity and physical size have entirely other limitations, though tidal turbines and stream generators that are being used are very similar to wind turbines.

It is effective at low speeds. Water is a thousand times more dense than air, which makes it possible to produce electricity at low speeds. Based on calculations, power can be generated even at 1 m per second.

It has a long lifespan. So far, there is no reason to believe that tidal energy plants are not long lived. This means an **ultimate** reduction of the money spent on selling the electricity, making this energy source a very cost-competitive one. As an example, the La Rance Tidal Barrage Power Plant was constructed in 1966 and is still generating large amounts of electricity up to this day.

It reduces foreign importation of fuel. By harnessing tidal energy on a large scale, we can help reduce foreign fuel importation and enhance energy security, as people would no longer have to rely much on foreign fuel imports to satisfy the growing energy demand.

It serves as coastal protection. Small dams and barrages, which are used to harness tidal energy, could protect ship ports and coastal areas from the dangerous tides during storms and bad weather conditions.

4.2　Disadvantages of tidal energy

It still has some environmental effects. As previously mentioned, tidal power plants are suspected to have some environmental effects, but these are yet to be determined. As we know, these facilities generate electricity with the use of tidal barrages that rely on ocean level manipulation, thus potentially having the same environmental effects as hydroelectric dams. Also, the turbine frames may potentially disrupt the natural movement of marine animals, and the construction of the whole plant may also disturb fish migration. Nevertheless, technological solutions are now being developed to resolve these issues.

It is an intermittent energy source. Tidal energy is considered as an intermittent source of energy, as it can only provide electricity when the tide surges, which happens about 10 h per day on average. This means that tidal energy can only be considered as reliable when accompanied with effective energy storage solutions.

It should be close to land. Tidal energy facilities need to be constructed close to land, which is also the place where technological solutions that come with them are being worked on. It is hoped that in a few years we will be able to use weaker tidal currents at locations further out to sea. In addition to this disadvantage, the areas where this energy is produced are far away from the exact locations where it is consumed or needed.

It is expensive. We should know that the method of generating tidal energy is relatively a new technology. It is projected that it will be commercially profitable by 2020 in larger scales with better technology. Also, the plants that harness this type of energy are linked to higher upfront costs required for construction. Thus, tidal energy displays a lack of cost effectiveness and efficiency in the world's energy markets and more technological advancements and innovations are still needed to make power commercially viable. To bring forward the greatest economical benefit to all concerned, it may be appropriate for some state aid to incentivize innovation and prevent unnecessary monetary wastage at a later stage.

It is still considered a new technology. Still a more theoretic source of power, tidal energy is limited in real life to just a few prototype projects because the technology has just begun to develop and needs plenty of research and huge funds before it reaches commercial status. There is a perceived lack of published fundamental marine turbine research and few publicly available agreed upon definitions of fundamental properties. Few standardized experts can provide full initial concept appraisal. Much worse, various developers are critical of those who utilize the best possible tidal sites to make their devices look more attractive than is either realistic or honest.

It requires long gestation time. The time and cost overruns can be huge for tidal power plants, which led to some of them being canceled, such as the UK's Severn Barrage. In fact, some tidal power stations, such as the one being planned in Russia, will never be realized because of a very long gestation time.

5. Difference between Tidal and Wave Energy

Tides and waves are two natural phenomena that occur on the surface of large water reservoirs like rivers, seas, oceans, and etc. Both of them are similar to each other on the basis that both are related to water systems but their potential in creating and transferring the energy differs a lot.

The tidal energy is harnessed by the rise and fall of sea levels. It occurs due to the gravitational pull of the Moon and Sun. There are two types of tidal energy: kinetic and potential energy. It can be converted to electrical energy using barrages, dams, tidal fences and tidal turbines. Tidal energy is far more reliable than other non-renewable resources since it is dependent on the gravitational pull of the Moon and Sun but is a discontinuous way as it is available for only 2 times a day. The engineering costs for harnessing the tidal energy are high but zero maintenance costs.

The wave energy is harnessed to waves moving on the surface of ocean due to wind. The strength of waves is under direct effect of wind strength. There are just two types of wave energy: kinetic energy and potential energy. It can be converted to electrical energy by installing special onshore and offshore systems. As compared to tidal energy, it is less reliable as it is dependent on the resultant effect of the wind speed but is a continuous way to obtain energy. It requires extremely high start-up costs to create and develop the technology required.

6. The Potential and Future of Tidal Energy

Worldwide, the tidal resources are considerable and also largely unmapped. However, global resources are estimated at 3 TW. The technically harvestable part of this resource, in areas close to the coast, is estimated by several sources at 1 TW. For an exact estimation of the actual resources, it is necessary to map the details per region or country. The shape of the coast also determines the tidal range with a fluctuating difference of up to 17 m between high and low tide. Argentina, Australia, Canada, Chile, China, Colombia, France, Japan, Russia, South Korea, Spain, the UK, and the US (Maine / Alaska) have very high tidal ranges. Furthermore, Eastern Africa has large resources for tidal range (see Fig. 7-1).

For tidal current technology, the stream speed needs to be at least 1. 5-2 m per second (m/s). The resources for this technology are very large, depending on the form and shape of the coast. They have not been mapped systematically worldwide. For Europe, the resources that are harvestable are estimated at a minimum of 12,000 MW (European Ocean Energy Association, 2010). Only a small number of countries,

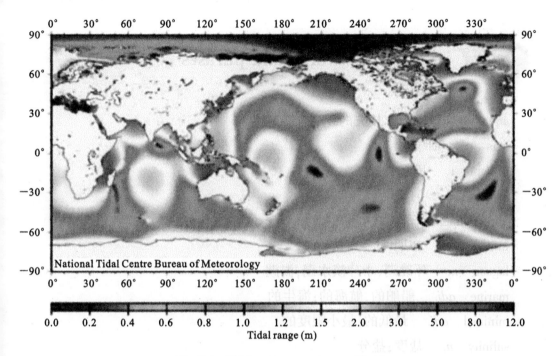

Fig. 7-1　Tidal range resources worldwide

Source: Bureau of Meteorology, Australian Government

e. g. , France, Ireland, Norway, the UK, the US, and some parts of the coast of China and Canada have been studied in detail. Other regions and countries with potential sources are Australia, Africa, India and Spain, but studies here are lacking.

An advantage of both tidal range and current energy is that they are highly predictable with daily, bi-weekly, biannual and even annual cycles over a longer time span of a number of years. Energy can be furthermore generated both day and night. Furthermore, tidal range is hardly influenced by weather conditions. Tidal stream is slightly more affected by the weather, but the fluctuations in the long run are lower than, for example, wind and solar. Another advantage of tidal stream energy is that the impact upon the landscape in the coastal zones is relatively small. Most structures are underwater and the associated requirements on land infrastructure can be relatively low or in some cases may be integrated into existing buildings or structures.

Source:

1. https://www. nationalgeographic. org/encyclopedia/tidal-energy.

2. https://www. sciencedirect. com/topics/engineering/tidal-energy.

3. https://differencebetweenz. com/difference-between-tidal-and-wave-energy.

新能源电力英语
English for Renewable Electricity Sources

New words and phrases

aquatic　*adj.*　与水相关的；水生的

axis　*n.*　轴线

disrupt　*v.*　破坏；使瓦解

estuary　*n.*　河口

impoundment　*n.*　蓄水；积水

infancy　*n.*　初期

lagoon　*n.*　环礁湖；小片淡水湖；潟湖

marine　*adj.*　船舶的，航海的；海生的

minimal　*adj.*　最低的；最小限度的

salinity　*n.*　盐度；盐分

silt　*n.*　淤泥；泥沙

submerge　*v.*　淹没；沉浸

suffocate　*v.*　阻碍；压制

surge　*n.*　大浪；波涛；波动

ultimate　*adj.*　最后的；最终的；根本的

Exercises

I. According to what you read in Text A, discuss the following questions with your partner.

1. What do you think of the cost indications of tidal energy? Please give your methods to reduce it.

2. What is the advantages of tidal energy in China?

3. Which is the major way to get tidal energy?

II. Fill in the blanks with the words and phrases in the text.

Tidal energy's source is a result 1 _____ the effects of the Sun and Moon's gravitational fields, combined 2 _____ our planet's rotation around its axis, 3 _____ results in low and high 4 _____. With this in mind, the power source of 5 _____ energy is potentially renewable, 6 _____ we are talking about tidal barrages, stream generators, 7 _____ the more recent technology, dynamic tidal power (DTP). Compared 8 _____ nuclear reserves and fossil fuels, the Sun and Moon's gravitational fields as well 9 _____ the Earth's rotation around its axis will not cease 10 _____ exist any time soon.

III. Please give the English or Chinese equivalents of the following technical terms.

1. tidal power
2. underwater land ownership
3. marine life
4. gestation time
5. the barrage gate
6. 潮汐潟湖
7. 淤积高度
8. 间歇性能源
9. 陆地基础设施
10. 激励创新

IV. Please translate the following sentences into English.

1. 在 20 世纪,工程师们开发出了利用潮汐运动来发电的方式。
2. 中国、法国、英国、加拿大和俄罗斯在使用这种能源方面有更大的潜力。
3. 目前有三种获得潮汐能的方式:潮汐流、拦河坝和潮汐潟湖。
4. 世界上第一座潮汐能发电站是 2007 年在北爱尔兰建设的。
5. 潮汐能受海平面升降影响,而波浪能是由于风引起的海浪在海面上的运动而产生的。

V. Please translate the following passage into Chinese.

Worldwide, the tidal resources are considerable and also largely unmapped. However, global resources are estimated at 3 TW. The technically harvestable part of this resource, in areas close to the coast, is estimated by several sources at 1 TW. For an exact estimation of the actual resources, it is necessary to map the details per region or country. The shape of the coast also determines the tidal range with a fluctuating difference of up to 17 m between high and low tide. Argentina, Australia, Canada, Chile, China, Colombia, France, Japan, Russia, South Korea, Spain, the UK, and the US (Maine / Alaska) have very high tidal ranges. Furthermore, Eastern Africa has large resources for tidal range.

Text B Drivers and Barriers of Tidal Energy

The potential of tidal energy is significant, particularly in certain locations, and the successful demonstrations of full-scale tidal current technologies in the last few years have mobilised the support of governments and private investment in the technology and project development in those regions.

The most important driver for tidal range and tidal current energy is that both technologies can generate renewable electricity close to urban centres, without becoming a **nuisance** or having a negative environmental impact on the landscape. Positively, tidal range **installation** can, while also contributing to or being part of water defences or **sluices**, have a **minimal** effect on the landscape, with **negligible** emissions and noise. However, there are a number of barriers that need to be overcome.

1. Technology Barriers

The technological challenge for tidal range is to increase the efficiency of the turbines. For tidal current technologies, the basic technologies exist but technical challenges continue to arise due to insufficient experience with materials, working and fixing structures in a harsh environment, demonstration, a lack of information and knowledge regarding performance, lifespan, operation and maintenance of technologies and power plants. For tidal current technology to become a real alternative to conventional energy sources, increased attention needs to be paid to technical risks in design, construction, installation and operation. According to reports of the Crown Estate (2013) and the Carbon Trust (2012), costs need to be brought down to at least 50%, which is comparable to offshore-wind energy generation costs. Moreover, importing knowledge and experience from other industry sectors, such as offshore oil and gas installations and offshore wind farms, including risk assessments, environmental impact assessments and engineering standards, is of great importance. This is not an easy process as much of this information is **proprietary** and of competitive advantage to firms. Furthermore, oil and gas technologies are often not the same as technologies for renewable projects (e. g. , high spec, high cost, one-off uses vs. lower cost, mass produced). More extensive research on new materials and

methodologies, and **rigorous** testing on new sub-components and complete functional prototypes is still necessary to establish these new technologies. For tidal current technologies, costs of fixtures to the seabed, and maintenance and installation costs need to be brought down. Furthermore, more experience in deploying **arrays** is required.

2. Ecological Impacts

The potential for traditional tidal range technology, which closes streams or river arms with dams or in **impoundments**, is limited due to ecological constraints. Additionally, experiences with artificially closed compounds have demonstrated that the costs of managing an artificial tidal basin (e. g. , in the case of La Rance and Cardiff Bay) are high and need careful monitoring and planning. The Canadian plants are noteworthy; there was a well-documented discussion from the start of the operation in these plants about the effects on fish and marine life and how to mitigate them. This information is currently of high value as ecological issues set important requirements and conditions for the permitting of installations in protected water bodies. On the other hand, the re-opening of dams and barriers, often built between the 1950s and 1970s can have great ecological benefits for the water bodies behind them due to a creation of a gradient that is beneficial to aquatic ecology (brackish water) and an increased oxygen content; in such instances, tidal technology can also be used as a tool for water quantity management, whilst generating power. A more innovative type of tidal range technology, which does not close impoundments completely, is currently in the developmental phase and will also be of interest. The challenge for tidal stream technologies is different. The ecological impacts are deemed to be less than tidal range technologies, but environmental regulators lack the appropriate expertise or tools to assess the environmental risks (Copping, 2013). Furthermore, baseline data of biodiversity in sea waters is limited, resulting in increased costs for evidence gathering and post deployment monitoring (Renewables UK, 2013).

3. Lack of Industrial Cohesion

The development of tidal stream technologies has been linked to small and micro enterprises, many of which have been spin-offs from university projects.

Consequently, there is a lack of cohesion within the industry, with many different designs and a number of small-scale producers. However, large turbine manufacturers such as ABB, Alstom, Andritz, Siemens, and Voith Hydro have entered this emerging sector by becoming involved in the start-up phase. This new interest is creating the conditions necessary to scale up the existing full-scale demonstration turbines into arrays. Since the full-fledged development and operating costs are still not clear, but can be expected to be high, especially during the start-up phase, the projects can become unviable for small and medium enterprises. Tidal energy still requires investment and R&D to develop and deploy viable and scalable commercial technology and infrastructure, better understand environmental impacts and benefits, and to achieve market entry. Most of the new projects are oriented towards helping bring technologies to a pre-commercial status, promoting easy access to research facilities or supporting the creation of new demonstration sites at sea. There remains a lack of knowledge of many different issues including those on various environmental impacts (e. g., mammal interaction or the impacts on the coastline due to tidal dissipation). Tidal energy technologies will require similar supply chains to offshore wind and oil and gas. The involvement of large and multi-disciplinary industries can be expected to promote synergies, which will generate economies of scale and reduce costs.

4. The Need for New Finance Mechanisms

Most project costs for tidal stream technologies are provided through government funds, or by technology developers themselves. Australia, Canada, France, Ireland (SEAI, 2010), South Korea (Hong, Shin and Hong 2010), and the UK have had active policies to support research and demonstration of tidal current technologies (IEA-OES, 2014c). Some countries promote a number of selected projects (e. g., in the Netherlands), while others have started a more active policy on marine energies (e. g., feed-in tariffs and requests for proposals in the Canada, France, and the UK). However, it is still difficult to provide the necessary financial framework conditions in the long term (beyond 2020). The need for new finance mechanisms is particularly relevant for the tidal stream technologies that have been tested at full scale, but will require market pull mechanisms to deploy at scale (Bucher and Couch, 2013).

Possible ways of attracting investments could be by offering tax rewards for investors, by attracting end-users, or by feed-in tariffs that would make high-cost, pre-commercial installations more attractive. Furthermore, suitable mechanisms for risk sharing or lowering insurance risks could reduce the overall project costs.

5. Insufficient Grid Infrastructure

For tidal stream technologies, grid connections to onshore grids can also be problematic. Some coastal countries, such as Portugal, the Netherlands, Norway, the South West of the UK and some regions of Spain, have high voltage transmission lines available close to shore, but many coastal regions, where the tidal energy resource is available, lack sufficient power transmission capacity to provide grid access for any significant amount of electricity. Equally, a number of open sea test centres have yet to establish grid connections. Similar problems have been identified for offshore wind. In Europe, the European Commission together with industry and Member States is supporting the development of an integrated offshore grid structure to deliver offshore wind to consumers, notably through the activities of the North Seas Countries Offshore Grid Initiative (NSCOGI). This takes into account the growth possibilities for offshore wind farms and defines options to build a European offshore grid. However, as its name suggests, this initiative particularly covers the North Sea, which is surrounded by large conurbations and industries. It will be harder to make a case for the less-populated Atlantic coasts which have the greatest potential for tidal energy. Nevertheless, taking into account the needs of tidal energy, as well as wind energy, developing joint projects can be more efficient than retro-fitting. Costs can be shared and the developments of hybrid or multi-platform solutions are encouraged. Port facilities will also be important for further development. Installation, operation and maintenance (O&M) of marine systems is expensive and even more if this will be performed in highly turbulent and changeable waters. In order to reduce time and cost on O&M, the alternative of unplugging the tidal energy converters from their offshore emplacement and performing the maintenance at a safe and more accessible port facility, is being considered as a real option. This, together with other **auxiliary** services would need the appropriate space and port facilities, making it necessary to consider the correct planning and management of infrastructure for the coastal areas where tidal energy represents a real energy alternative. Furthermore, a number of open sea test centres have not established grid connections yet.

6. Planning and Licensing Procedures

Coastal communities and those engaged in more traditional marine activities tend to be critical of the impact of new, innovative technology. Planning and licensing processes for ocean energy therefore need to be open and comprehensive enough to take these concerns into account. However, in contrast to spatial planning on land, countries generally have limited experience with, and sometimes inadequate governance and rules for, planning and licensing in the marine environment. This is particularly true for sensitive areas in relation to environmental protection and nature conservation. The lack of processes for guidance, planning and licensing marine activities in areas where many different interests (transport, energy, tourism, fisheries, and etc.) coincide, tends to increase uncertainty and therefore a risk of delays or failure in marine projects. This can be a barrier to securing investments. Early and adequate involvement of stakeholders is also important under these circumstances. The challenge, for particularly innovative tidal energy projects, is to develop plans, which from the start significantly mitigate any negative environmental effects.

7. Integrating the Technology with Other Economic and Societal Functions

Given that tidal range technology is relatively new, most of the projects and work are particularly focused on the technology of the device itself and its direct infrastructure. However, for larger schemes, a connection to other factors such as shipping, recreation, water defence and ecological impact could not only bring down installation costs (by better coordination of infrastructure), but also other types of costs and societal acceptance. This is in part demonstrated through the Norwegian Road Administration (2012). For tidal stream technologies, there are a number of plans for hybrid systems combining floating offshore wind with tidal current technologies (e. g. , a 500 kW demonstration plant by MODEC in Japan). However, in most cases tidal current technologies do not match well with offshore wind parks as the required strong tides increase the installation costs of the offshore wind parks.

New words and phrases

array   *n.*   数组;排列;阵列

auxiliary   *adj.*   备用的;辅助的;协助的

impoundment   *n.*   蓄水;扣留

installation   *n.*   安装;装置

minimal   *adj.*   最小的;最低的

negligible   *adj.*   可以忽略的;微不足道

nuisance   *n.*   讨厌的人;麻烦事

proprietary   *adj.*   所有的;私人拥有的;专利的

rigorous   *adj.*   严格的;严厉的

sluice   *n.*   水闸;蓄水

Exercises

Discuss the following questions with your partners.

1. List the barriers of tidal energy that the writer mentioned in the text and give your practical solutions to overcome them.

2. Why is the potential for traditional tidal technology limited? Give your reasons.

Keys to Exercises

Unit 8
MHD

Text A   MHD

When we talked about conventional thermal power plants, the efficiency is in between 37% to 40%. In thermal plants, the fuel is burnt and the heat generated is used to convert water into steam. This steam is used to rotate turbine of an alternator. This is how steam energy is converted into mechanical energy and later mechanical energy is converted into electrical energy. Since in this process, there are many conversion from one form of energy to another, this results in losses and due to this the overall efficiency of these thermal power plant decreases.

1. What Is MHD?

1.1   The fundamental concept

Imagine a system where a heat energy is directly converted to electrical energy. The main advantage of such type of system is that higher efficiency is possible since there is no conversion of energy except heat energy to electrical energy. The direct conversion of heat energy into electrical energy is known as the **magneto hydrodynamic** (MHD) generation. The field of MHD was initiated by Hannes Alfvén[1]*, for which he received the Nobel Prize in physics in 1970.

MHD is the **physical-mathematical framework** that concerns the dynamics of magnetic fields in electrically **conducting fluids**, e. g. , in plasmas and liquid metals. The word magnetohydrodynamics is comprised of the words magneto — meaning magnetic, hydro — meaning water (or liquid) and dynamics referring to the movement of an object by forces.

MHD power generation is **elegantly** simple technique. Magneto hydro dynamics (magneto-fluid dynamics or hydro-magnetics) is the **academic discipline** which studies the dynamics of electrically conducting fluids. Examples of such fluids include plasmas, liquid metals, and salt water. The generator used in this process is called magneto hydro dynamic generator. It resembles the rocket engine surrounded by enormous magnet. It has no moving parts and the actual conductors are replaced by ionized gas (plasma). Hence it has a very high efficiency. Although the cost cannot be predicted very accurately, it has been reported that capital cost of MHD plants will be competitive with those of conventional steam plants.

The set of equations that describe MHD are a combination of the **Navier-Stokes** equations of **fluid dynamics** and Maxwell's equations of electromagnetism. These differential equations must be solved simultaneously, either analytically or numerically.

1.2 Structures in MHD

Watch Figure 8-1 to see the structures in MHD.

The simplest form of MHD, ideal MHD, **assumes** that the fluid has so little **resistivity** that it can be treated as a perfect conductor. This is the limit of infinite magnetic Reynolds number. In ideal MHD, Lenz's law dictates that the fluid is in a sense tied to the magnetic field lines. To explain, in ideal MHD a small rope-like volume of fluid surrounding a field line will continue to lie along a magnetic field line, even as it is twisted and distorted by fluid flows in the system. This is sometimes referred to as the magnetic field lines being "frozen" in the fluid. The connection between magnetic field lines and fluid in ideal MHD fixes the **topology** of the magnetic field in the fluid — for example, if a set of magnetic field lines are tied into a knot,

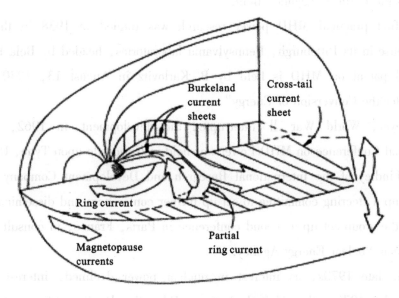

Fig. 8-1　Structures in MHD

then they will remain so as long as the fluid / plasma has negligible resistivity. This difficulty in reconnecting magnetic field lines makes it possible to store energy by moving the fluid or the source of the magnetic field. The energy can then become available if the conditions for ideal MHD break down, allowing magnetic reconnection that releases the stored energy from the magnetic field.

In many MHD systems most of the electric current is compressed into thin **nearly-two-dimensional ribbons** termed current sheets. These can divide the fluid into magnetic **domains**, inside of which the currents are relatively weak.

Another example is in the **Earth's magnetosphere**, where current sheets separate topologically distinct domains, isolating most of the Earth's ionosphere from the solar wind.

2. The History and Development of MHD

2.1　The history of MHD

The concept of MHD power generation was introduced for the very first time by Michael Faraday[2*] in the year 1832 in his Bakerian lecture to the Royal Society. However, the actual utilization of this concept remained unthinkable. In fact, he carried out experiments at Waterloo Bridge, measuring current from the flow of the

Thames in the Earth's magnetic field.

The first practical MHD power research was funded in 1938 in the US by Westinghouse in its Pittsburgh, Pennsylvania laboratories, headed by Bela Karlovitz. The initial patent on MHD is held by B. Karlovitz in August 13, 1940, naming "Process for the Conversion of Energy".

However, World War II interrupted the development. In 1962, the First International Conference on MHD Power was held in Newcastle upon Tyne, UK by Dr. Brian C. Lindley of the International Research and Development Company Ltd. The group set up a steering committee to set up further conferences and disseminate ideas. In 1964, the group set up a second conference in Paris, France, in consultation with the European Nuclear Energy Agency.

In the late 1970s, as interest in nuclear power declined, interest in MHD increased. In 1975, the United Nations Educational, Scientific and Cultural Organization (UNESCO) became persuaded the MHD might be the most efficient way to utilise world coal reserves, and in 1976, sponsored the ILG-MHD. In 1976, it became clear that no nuclear reactor in the next 25 years would use MHD, so the IAEA and ENEA (both nuclear agencies) withdrew support from the ILG-MHD, leaving UNESCO as the primary sponsor of the ILG-MHD.

The first recorded use of the word magnetohydrodynamics is by Hannes Alfvén in 1942:

"At last some remarks are made about the transfer of momentum from the Sun to the planets, which is fundamental to the theory. The importance of the magneto-hydrodynamic waves in this respect are pointed out."

The Japanese program in the late 1980s concentrated on closed-cycle MHD. In 1986, Professor Hugo Karl Messerle at the University of Sydney researched coal-fueled MHD. The Italian program began in 1989 with a budget of about 20 million USD, and had three main development areas: MHD modeling, superconducting magnet development and retrofits to natural gas power plants.

In space science. The magnetosphere is the outermost layer of the **geospace**, and the interaction of the solar wind with the magnetosphere is the key element of the space weather cause-and-effect chain process from the Sun to the Earth, which is one of the most challenging scientific problems in the geospace weather study. The **nonlinearity**, multiple component, and time-dependent nature of the geospace make it very difficult to describe the physical process in geospace using traditional analytic analysis approach. **Numerical simulations**, a new research tool developed in recent decades, have a deep impact on the theory and application of the geospace. MHD simulations started at the end of the 1970s, and the initial study was limited to two-dimensional (2D) cases. Due to the intrinsic three-dimensional (3D) characteristics of the geospace, 3D MHD simulations emerged in the 1980s, in an attempt to model the large-scale structures and fundamental physical processes in the magnetosphere. They started to combine with the space exploration missions in the 1990s and make comparisons with observations. Physics-based space weather forecast models started to be developed in the 21st century. Currently only a few space-power countries such as USA and Japan have developed 3D magnetospheric MHD models. Figure 8-2 shows the MHD simulation of the solar wind.

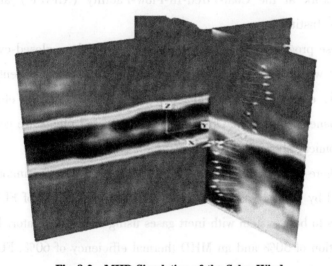

Fig. 8-2   MHD Simulation of the Solar Wind

In the recent years, with the rapid advance of space science in China, we have developed a new global MHD model, namely PPMLR-MHD, which has high order spatial accuracy and low numerical dissipation. At present, we will briefly introduce the global 3D MHD modeling, especially the PPMLR-MHD code, and summarize our recent work based on the PPMLR-MHD model, with an emphasis on the interaction of interplanetary shocks with the magnetosphere, large-scale current systems, reconnection voltage and transpolar potential drop, and Kelvin-Helmholtz (K-H) instability at the **magnetopause**.

2.2　The distribution of MHD development in the world

The MHD development isn't balanced in the world. In advanced countries, the power generation with MHD generators is already in use but in developing countries it's still under construction.

(1)Development in some countries.

Over more than a ten-year span, Bosnian engineers in Bosnia, in the Institute of Thermal and Nuclear Technology, EnergoInvest Co., Sarajevo[3*], had built the first experimental Magneto-Hydrodynamic facility power generator in 1989. It was the first patented.

In the 1980s, the US Department of Energy began a vigorous **multiyear** program, culminating in a 1992 50-MW demonstration coal at the Component Development and Integration Facility (CDIF) in Butte, Montana. This program also had significant work at the Coal-Fired-In-Flow-Facility (CFIFF) at University of Tennessee Space Institute.

The Japanese program in the late 1980s concentrated on closed-cycle MHD. The belief was that it would have higher efficiencies, and smaller equipment, especially in the clean, small, economical plant capacities near 100 megawatts (electrical) which are suited to Japanese conditions. Open-cycle coal-powered plants are generally thought to become economical above 200 megawatts.

In 1994, there were detailed plans for FUJI-2, a 5-MW continuous closed-cycle facility, powered by natural gas, to be built using the experience of FUJI-1. The basic MHD design was to be a system with inert gases using a disk generator. The aim was an **enthalpy** extraction of 30% and an MHD thermal efficiency of 60%. FUJI-2 was to be followed by a retrofit to a 300-MW (electrical) natural gas plant.

The Italian program began in 1989 with a budget of about 20 million USD, and had three main development areas: MHD modelling, superconducting magnet development, and retrofits to natural gas power plants. One was to be in Ravenna[4*]. In this plant, the combustion gases from the MHD would pass to the boiler. The other was a 230-MW (thermal) installation for a power station in Brindisi[5*] that would pass steam to the main power plant.

(2)MHD energy development and application in China in recent years.

In recent years, the idea of green development has prevailed in global energy governance. It is now an international consensus that energy cooperation should aim at the development of clean energy.

On part of China, it is now the world's largest energy producer and consumer. In energy conservation, emission reduction and renewable energy development, China's achievements over the years have been widely recognized.

At present, China has greater installed capacity in MHD, hydropower, wind power and solar power than any other country in the world. Of the total energy consumption in 2016 (4. 36 billion tons of standard coal), non-fossil fuel accounted for 13. 3%, which was 1. 3 percentage points higher than the previous year. In the first three quarters of 2015, GDP per unit of energy use downed by 5. 2% year on year. Such progress is largely attributed to green development.

On September 4th, 2016, important outcomes on the issue of energy were reached at the **G20 Hangzhou Summit**. By highlighting the importance of efficient and clean energy in the future, the G20 demonstrated its unique role in global energy governance. MHD and other clean energy will be used widely.

By working together, all of the countries will make even greater contribution to green, inclusive and sustainable development in building a community of shared future for mankind.

3. How Is MHD Generated?

3.1　Principle of MHD generation

The principal of MHD power generation is very simple and is based on Faraday's law of electromagnetic induction, which states that when a conductor and a magnetic field moves relative to each other, then voltage is induced in the conductor, which results in flow of current across the terminals. Here conductor can be in solid, liquid or gaseous form.

As the name implies, the magneto hydrodynamics generator, is concerned with the flow of a conducting fluid in the presence of magnetic and electric fields. In conventional generator or alternator, the conductor consists of copper windings or strips while in an MHD generator the hot ionized gas or conducting fluid replaces the solid conductor.

A pressurized, electrically conducting fluid flows through a transverse magnetic field in a channel or duct. Pair of **electrodes** are located on the channel walls at right angle to the magnetic field and connected through an external circuit to deliver power to a load connected to it. Electrodes in the MHD generator perform the same function as brushes in a conventional DC generator. The MHD generator develops DC power and the conversion to AC is done using an inverter.

The power generated per unit length by MHD generator is approximately given by,

$$P = \frac{\sigma u B^2}{P}$$

Where, u is the fluid velocity, B is the magnetic flux density, σ is the electrical conductivity of conducting fluid and P is the density of the fluid.

It is evident from the equation above, that for the higher power density of an MHD generator there must be a strong magnetic field of 4-5 **tesla** and high flow velocity of conducting fluid besides **adequate** conductivity.

3.2   The construction of MHD generation

According to the principle of electromagnetic induction, a conductive fluid (gas or liquid) is generated relative to the magnetic field.

Magnetic fluid power generation is divided into open cycle system, closed cycle system and liquid metal circulation system according to the cycle of working fluids. The simplest open magnetic flux generator consists of a **combustion chamber**, a power generation channel and a magnet. The working process is: in high temperature, gas produced after the combustion of fuel, the easily **ionized potassium** or sodium salt is added to make it partial ionization and the nozzle is accelerated to produce a high temperature and high speed conductive gas (partial plasma) with a temperature of 3,000 ℃ and a speed of 1,000 meters per second.

3.3   MHD cycles and working fluids

The MHD cycles can be of two types, namely open cycle MHD and closed cycle MHD. The detailed account of the types of MHD cycles and the working fluids used, are given below.

The MHD generation or, also known as magneto hydrodynamic power generation is a direct energy conversion system which converts the heat energy directly into electrical energy, without any intermediate mechanical energy conversion, as opposed to the case in all other power generating plants. Therefore, in this process, substantial fuel economy can be achieved due to the elimination of the link process of producing mechanical energy and then again conversion. The hot ionizing gases at a temperature of about 2,500 ℃ are passed through the MHD duct across a strong magnetic field which is at an angle of 90° to the flow of the conducting fluid or gases (called plasma). The **ionisation** of gas is produced either by thermal means, i. e., at an elevated temperature or by seeding with a substance like cesium or potassium vapours which ionizes at relatively low temperatures. The atoms of seed element split off electrons. The presence of the negatively charged electrons makes the gas an electrical conductor.

Due to hot and ionization nature of gases, they behave as the conductor and hence an EMF is induced across the electrodes as shown in Fig. 8-3. When the electrodes are connected to the load, current starts flowing through the load.

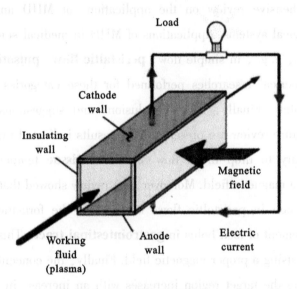

Fig. 8-3   Closed cycle MHD system

This system is usually advantageous if the MHD system is used along with thermal plants that utilize the heat of the exhaust gases leaving the MHD ducts. Due to this combination, the efficiency of a thermal power plant can be raised to 60%.

The study of MHD flow is significant due to its many engineering applications such as the cooling of reactors, **electrostatic precipitation**, power generators, MHD pumps, **accelerators**, petroleum industry, and the design of heat exchangers.

4. The Application of MHD

MHD is used widely in many fields at present.

In the early 60s of last century, China began to study magnetic fluid power generation, and built test bases in Beijing, Shanghai and Nanjing. According to the rich characteristics of China's coal resources, China will focus on the research of coal-fired magnetic fluid power generation, and take it as one of the two research topics in the "863" energy field, and **strive** to catch up with the world's advanced level in a short time.

The MHD applications in medical sciences.

MHD fluid flow in different geometries relevant to human body parts is an interesting and important scientific area due to its applications in medical science. We perform a comprehensive review on the applications of MHD and their numerical modelling in biological systems. Applications of MHD in medical science are classified into four categories, e. g. , in simple flow, **peristaltic flow**, **pulsatile flow**, and drag delivery. The numerical researches performed for these categories are reviewed and summarized separately. Finally, some conclusions and suggestions for future works based on the literature review are presented. The results indicated that during a surgery when it is necessary to drop blood flow or reduce **tissue** temperature, it may be achieved by using a magnetic field. Moreover, the review showed that the trapping is an important phenomenon in peristaltic flows that causes the formation of thrombus in blood and the movement of food bolus in **gastrointestinal tract**. This phenomenon may be disappeared by using a proper magnetic field. Finally, the concentration of **particles** that are delivered to the target region increases with an increase in the magnetic field intensity.

MHD effects arising in plasma sources, such as short-pulsed magneto-plasma-dynamic generators and inductively heated plasma generators, are analyzed with both algebraic models and measured data. Functional principles of the sources based on their MHD behavior are explained. Moreover, Stewart numbers in the order of magnitude of at least 10^{-1} are calculated for the systems and qualified as an **identifier** for the magnetic influence on plasmas. Here, the considered plasma systems that are a priori known as MHD systems are used to determine typical values of the Stewart number.

Based on this experience the concept of a **plasma probe** to magnetically influence or control weakly ionized free stream plasma flows is presented.

The MHD is used in **angular rate sensor**. The MHD angular rate sensor (MHD ARS) with the characteristics of light weight, small size, and broad bandwidth is a perfect inertia sensor for space application.

Application of the MHD in Marine Vessels. In some MHD applications, the electric current is applied to create MHD propulsion force. An electric current is passed through seawater in the presence of an intense magnetic field. Functionally, the seawater is then the moving, conductive part of an electric motor, pushing the water out the back and accelerating the ship.

In a word, the application of MHD will be much wider in the near future.

5. The Advantages and Disadvantages of MHD

The advantages and disadvantages of renewable energy show us that developing this technology is important for the generations to come. Global warming is a potential threat. Pollution is a potential threat. Whether one believes they are man-made issues or part of the natural cycle of the planet, the end result is the worst. The main advantage of MHD system is that higher efficiency since there is no conversion of energy except heat energy to electrical energy. The direct conversion of heat energy into electrical energy is known as the magneto hydrodynamic (MHD) generation. But there are also disadvantages of the MHD energy.

5. 1   The advantages of MHD generation

The advantages of MHD generation over the other conventional methods of generation are given below.

(1) Free of pollution. It contributes greatly to the solution of serious air and thermal pollution faced by steam plants. In MHD, the thermal pollution of water is **eliminated** through clean energy system. The temperature of working fluid is maintained by the walls of MHD.

(2) Use of MHD plant operating in conjunction with a gas turbine power plant might not require to reject any heat to cooling water.

(3) Less complicated than the conventional generators, having simple technology.

(4) There are no moving parts in generator which reduces the energy loss. Since it has no moving parts, so it is more reliable. It has the ability to reach the full power level as soon as it is started. A large amount of power is generated.

(5) These plants have the potential to raise the conversion efficiency up to 55%-60%. Since conductivity of plasma is very high (can be treated as infinite), it contributes to better utilisation of fuel.

(6) The size of the plant is considerably smaller than conventional fossil fuel plants.

(7) It is applicable with all kind of heat source like nuclear, thermal, thermonuclear plants and etc. Extensive use of MHD can help in better fuel utilization.

(8) The price of MHD generators is much lower than conventional generators. It's suitable for a peak power generation.

5.2 The disadvantages of MHD generation

There are also disadvantages of MHD generation.

(1) The construction of superconducting magnets for small MHD plants of more than 1 kW electrical capacity is only on the drawing board.

(2) Difficulties may arise from the exposure of metal surface to the intense heat of the generator and form the **corrosion** of metals and electrodes.

(3) Construction of generator is uneconomical due to its high cost. Construction of heat resistant and non conducting ducts of generator and large superconducting magnets is difficult.

(4) MHD without superconducting magnets is less efficient when compared with combined gas cycle turbine.

In a word, there are advantages and disadvantages of MHD energy. But developing this technology is important for the generations to come.

6. The Future Development of MHD

The future power generation with MHD generators will be more and more important. It can therefore be claimed that the development of MHD for electric utility power generation is an objective of national significance. The practical efficiency of this type of power generation will not be less than 60% in the future.

Since a decade the demand for electricity is increasing at alarming rate and the demand for power is running ahead of supply. The present day, methods of power generation are not much efficient and it may not be sufficient or suitable to keep pace with ever increasing demand. The recent severe energy crisis has forced the world to rethink and develop the magneto hydrodynamic (MHD) type power generation which remained unthinkable for several years after its discovery. It is a unique and highly efficient method of power generation with nearly zero pollution. Because it is the generation of electric power directly from thermal energy utilizing the high temperature conducting plasma moving through an intense magnetic field.

In advanced countries this technique is already in use; but in developing countries, it's still under construction. Efficiency matters the most for establishing a power plant. MHD power plants have an overall efficiency of 55%-60%, but it can be **boosted up** to 80% or more by using superconducting magnets in this process. Whereas the other non-conventional methods of power generation such as solar, wind, geothermal, tidal have a highest efficiency not more than 35%. Hence by using MHD power generation method separately or by combined operation with thermal or nuclear plants, we hope to bring down the energy crisis at a high rate.

Indeed, in China, as **urbanization** and agricultural modernization remains an ongoing process, energy conservation and emission reduction becomes a daunting task. One example is the autumn and winter haze resulting from air pollution.

China is taking on this task. We will strive to make the growth in energy supply mainly green and low-carbon. To this end, we have set a number of targets to be met by 2020. These include:

Increasing the share of non-fossil fuel in primary energy consumption to 15%;

Increasing the proportion of natural gas to at least 10%;

Keeping the percentage of coal consumption below 58%.

This will enable us to reach the **emissions peak** before 2030.

In conclusion, future power generation with MHD generators can make rapid **commercialization** possible with the improvement in corrosion science and superconducting magnets, saving billions of dollars towards fuel prospects of much better fuel utilization.

It can therefore be claimed that the development of MHD for electric utility power generation is an objective of national significance.

The practical efficiency of this type of power generation will not be less than 60%. Hence it will be most significant in upcoming decade.

If we solve this (making MHD cost effective), we will succeed, otherwise MHD will be in proceedings and papers.

Source:

1. https://www. iea. org/fuels-and-technologies/nuclear.

2. https://www. britannica. com/science/nuclear-fission.

3. https://core. ac. uk/reader/5222143.

4. http://www. iccf11. org/green-energy.

5. https://sciencing. com.

6. https://www. iea. org/reports/ renewr-power-in-a-clean-energy-system.

New words and phrases

accelerator   *n.*   加速器

adequate   *adj.*   足够的

assume   *v.*   假定;认为

boost up   提升;提高

combustor   *n.*   (喷射引擎的)燃烧室;燃烧装置

commercialization   *n.*   商业化;商品化

corrosion   *v.*   腐蚀;侵蚀;锈蚀

domain   *n.*   范围

electrode   *n.*   电极;电焊条

elegantly   *adv.*   优美地;雅致地;高雅地

eliminate   *v.*   排除;消除;淘汰

enthalpy   *n.*   焓;热函;热含量

geospace   *n.*   地球空间

identifier   *n.*   检验人;标识符

ionisation　*n.*　离子化;电离

magnetopause　*n.*　磁层顶

multiyear　*n.*　多年

nonlinearity　*n.*　非线性

particle　*n.*　微粒

resistivity　*n.*　抵抗力;电阻系数

strive　*v.*　努力奋斗;力求

tesla　*n.*　特斯拉(磁通量密度单位)

tissue　*n.*　(动植物的)组织

topology　*n.*　拓扑学;拓扑结构

urbanization　*n.*　都市化

academic discipline　学科;学术领域

angular rate sensor　角速率传感器

combustion chamber　燃烧室

conducting fluid　导电液体

Earth's magnetosphere　地球磁层

electromagnetic induction　电磁感应

electrostatic precipitation　静电沉淀

emission peak　排放量峰值

fluid dynamics　流体动力学

G20 Hangzhou Summit　20 国集团杭州峰会

gastrointestinal tract　胃肠道

ionized potassium　电离钾

MHD Modeling　MHD 建模

magneto hydrodynamic　磁流体动力学

Navier-Stokes equations　纳维-斯托克斯方程

nearly-two-dimensional ribbon　近二维色带

numerical simulation　数值模拟

peristaltic flow　蠕动流

physical-mathematical framework　物理-数学框架

plasma probe　等离子(区)探测器

pulsatile flow　脉动流

solar wind　太阳风

superconducting magnet　超导磁体

Notes

1* Hannes Alfvén　汉勒斯·阿尔文,瑞典天文学家,因在磁流体动力学方面的基础性研究和发现等,于1970年获诺贝尔物理学奖。

2* Michael Faraday　迈克尔·法拉第,英国物理学家、化学家,发现了电磁感应现象并提出力场概念。

3* Sarajevo　萨拉热窝,南斯拉夫中部城市。

4* Ravenna　拉文那,意大利东北部港市。

5* Brindisi　布林迪斯,意大利东南部城市。

Exercises

Ⅰ. **Read the text and discuss over the following questions with your partner.**

1. What is MHD? Please describe it with your partners.

2. What do you think is the future of MHD in China?

3. What are the advantages of MHD? Give some examples to describe the MHD applications.

Ⅱ. **Please give the Chinese or English equivalents of the following terms.**

1. 导电流体的磁性

2. 地球磁层

3. 磁流体动力学发电建设

4. 电磁感应原理

5. 在医学中的运用

6. 物理-数学框架

7. 流体动力学

8. 近二维色带

9. electrostatic precipitation

10. emission peak

III. Please translate the following sentences into English.

1. 这一定律表明,在磁场中运动的导体中会产生电动势。

2. 顾名思义,磁流体动力学发生器是指在磁场和电场存在下,导电流体的流动。

3. 由于这种结合,火力发电厂的效率可以提高 60%。

4. 在发达国家,MHD 发电机已投入使用;但在发展中国家,它仍在建设中。

5. 近年来,绿色发展理念在全球能源治理中盛行。

6 通过共同努力,所有国家将为绿色、包容和可持续发展做出更大的贡献,从而为人类建设一个未来的共同体。

7. 下面给出了磁流体发电相对于其他传统发电方式的优点。

8. 因此,可以说,开发用于电力发电的 MHD 是一个具有全国性意义的目标。

9. 事实上,在中国,城市化和农业现代化仍在进行中,节能减排成为一项艰巨的任务。

10. 因此,在今后十年中,它将是最重要的。

IV. Please translate the following passage into Chinese.

An MHD generator is an expansion engine. It delivers electric rather than mechanical power. There are virtually no upper limits to the temperature it can tolerate, the rapidity of its response. There are no upper limit to the power a single unit can be designed to deliver, either.

The behavior of the plasma is uncomplicated compared to that encountered in some other devices. But the behavior of the plasma is complex, because of the precision with which one needs to know it. The topics included are: generator configurations; electron density; electron mobility. They also include mixture rules; the Hall effect; uniformity; two-temperature plasma and ionization growth at a channel inlet; ionization instability; high enthalpy extraction experiments. They also include segmenting; electrode voltage drop; arcs and electrodes; electrical effects of slag; current control and waves.

Text B   The MHD System

The MHD generator needs a high-temperature gas source, which could be the **coolant** from a nuclear reactor or more likely high-temperature combustion gases generated by burning fossil fuels, including coal, in a combustion chamber. There are some different types of MHD systems depending on the different category.

1. Classification by Structure

There are two types of magneto hydrodynamic generation systems according to the structure — the open- and closed-cycle MHD system.

1.1   Open-cycle MHD systems

Watch the picture of the open-cycle MHD systems (see Fig. 8-4).

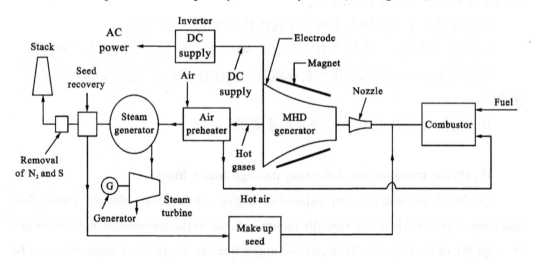

Fig. 8-4   Open-cycle MHD systems

· Working fluid after generating electrical energy is discharged to the atmosphere through a stack.

· Operation of MHD generator is done directly on combustion products.

· Temperature should be at 2,300 ℃ to 2,700 ℃.

· Open-cycle system is more developed.

An elementary open-cycle MHD system is a system in which a high-pressure, high-temperature combustion gas is forced through a strong magnetic field.

Coal is processed and burnt in the **combustor** at a high temperature of about 2,600 ℃ and pressure of about 12 atmospheres with pre-heated air to form the plasma. Then a seeding material, such as **potassium carbonate**, is injected to the plasma in order to increase the electrical **conductivity**.

The resulting mixture having an electrical conductivity of about 10 S/m^{1*}, is expanded through a nozzle, so as to have high velocity, and then passed through the strong magnetic field (5-7 T) of the MHD generator. During the expansion of gas at high temperature, **the positive and negative ions** move to the electrodes and so constitute an electric current. This current is DC and an inverter is employed for its conversion into AC.

The gas leaving the MHD generator is still very hot. The heat from the exhaust gases of the MHD generator is utilised in preheating the air supplied to the combustor. The seed material is recovered from the gas for successive use and harmful emissions (such as nitrogen and sulphur) are removed from the gas, for pollution control, and the gas is finally discharged to the atmosphere through a stack.

The open-cycle MHD system explained above is not suitable for commercial use. For making this process efficient, it is necessary to combine the MHD unit with **steam turbine-alternator unit**. In this system, the heat from the **exhaust gases** of the MHD generator is used to raise steam which generates additional energy in a steam turbine-alternator unit. A part of this steam is also used in a steam turbine driving a compressor for compressing air for the MHD cycle. Such a cycle is called the hybrid, binary or topping MHD steam plant cycle.

The electrodes are usually made of **graphite** and **the duct of boron nitride**.

Any type of fossil fuel (coal, oil, natural gas) can be used in MHD generator but a direct coal fired MHD generator has the following advantages.

① Slag from coal combustion coats the generator electrodes and protects the generator from electrical and mechanical corrosion that otherwise occurs in a clean fuel generator. However, the electrodes may be short circuited by molten ash.

② Coal contains less hydrogen and, therefore, the sink for electrons in the flow created by the presence of OH ions is reduced.

Char, having almost no hydrogen, is better than coal even. It is easier to handle and feed in comparison to coal. It results in a 25% increase in the performance of the

generator.

Power generated by the MHD system is given as:

$$\rho = \sigma B2v2K (1 - K) W/m^3$$

Where σ is the specific **electrical conductivity** of gas in S/m, B is **magnetic field strength** in Tesla (Wb/m^2), v is the velocity of gas in m/s, and K is the **ratio** of **external load voltage** to open-circuit voltage.

1.2　Closed-cycle MHD systems

· Helium or argon (with cesium seeding) is used as the working fluid.

· Working fluid is recycled to the heat sources and is used again.

· Temperature should be at about 1400 ℃.

· Closed-cycle system is less developed.

As the name suggests, the working fluid, in a closed cycle, is circulated in a closed loop. The closed-cycle MHD systems (see Fig. 8-5) may be either a plasma converter or a liquid metal converter. The plasma converter uses an ionized gas (helium or argon seeded with cesium) and the liquid metal converter uses the vapour of the metal or the metal in a liquid form (the metal may be an alkali or some other metal).

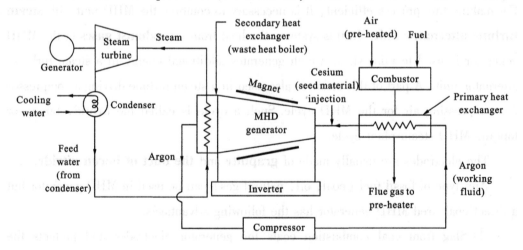

Fig. 8-5 Closed-cycle MHD systems

Liquid metal system has the basic advantage of high electrical conductivity. So high temperatures are not required in this system to achieve the high electrical conductivity and the systems are normally designed to operate at temperatures below 1,400 ℃ (a much lower temperature in comparison to plasma converters). This temperature is low enough that the energy can be supplied by a nuclear reactor or fossil fueled system.

The complete system can be divided into three distinct but interlocking loops. Coal is gasified and the gas so produced has heat value of about 5.35 MJ/kg and a temperature of about 520 ℃. This gas is burnt in a combustor to generate heat. This heat is transferred to the working fluid (argon) of the MHD cycle in a heat exchanger (HX1). The combustion products are passed through the air preheater (for recovery of a part of heat of combustion products) and purifiers and then discharged to the atmosphere. This is loop I and may be called an external heating loop.

The hot argon gas is seeded with cesium and passed through the MHD generator that produces direct current (DC). This DC output is converted into AC by means of inverter and then supplied to the grid. This is the loop II and may be called an MHD loop.

For further recovery of heat from the working fluid and its use in generating of steam, the working fluid is slowed down, in a diffuser, to a low subsonic speed and then this fluid is passed through the heat exchanger (HX2). In the heat exchanger (HX2), the fluid imparts its heat to water and so generates steam.

This steam is used partly for driving a steam turbine operating the compressor and partly expanded in a steam turbine driving a three-phase alternator. The working fluid is then passed through compressor and **intercooler** and then returned back to the heat exchanger (HX1). This is the loop III and may be called a steam loop.

The super heated metallic vapour is expanded through a supersonic nozzle into the drift tube or mixer. **Atomized subcooled liquid droplets** are accelerated by the vapour. The vapour also condenses on the liquid droplets so that the fluid entering the MHD generator is essentially a **liquid**. The resulting velocity of the fluid is more than 150 meters per second.

The advantages of liquid metal system are as follows.

① This system can use nuclear energy as high temperature is not the requirement of this system as in case of the plasma converter.

② It can easily provide AC power supply directly, while it is almost impossible to do so in a plasma system.

③ The size of the system including that of magnets is comparatively smaller. This is because of high power density.

However, the liquid metal system has the following limitations.

① The metallic vapours are poor electrical conductors.

② High velocities cannot be obtained by expansion in this system while it is much easier to achieve a high fluid velocity employing a gas and a nozzle. This is because the liquids are practically **incompressible**.

③ **The overall conversion efficiencies obtainable** with liquid metal system are quite below to that of plasma system.

A closed-cycle system can provide more useful power conversion at lower temperatures of about 1,600 ℃, but this system has not **taken** practical **shape** so far. The difficulties with such a system are the design of heat exchanger (the heat exchanger operates up to the highest temperature of the gas), requirement of absolute purity of the working fluid, the problems posed by electrical stability of the flow in the generator (the gas is subject to electric fields approaching breakdown conditions) and etc.

2. Classification by Source of Driving Force

2.1　Coal-fired MHD systems

The choice of type of MHD generator depends on the fuel to be used and the application. The **abundance** of coal reserves throughout much of the world has favoured the development of coal-fired MHD systems for electric power production. Coal can be burned at a temperature high enough to provide **thermal ionization**. However, as the gas expands along the duct or channel, its electrical conductivity drops along with its temperature. Thus, power production with thermal ionization is essentially finished when the temperature falls to about 2,500 K (about 2,200 ℃, or 4,000 ℉). To be economically competitive, a coal-fired power station would have to combine an MHD generator with a conventional steam plant in what is termed a binary cycle. The hot gas is first passed through the MHD generator (a process known as topping) and then on to the **turbogenerator** of a conventional steam plant (the bottoming phase). An MHD power plant employing such an arrangement is known as an open-cycle, or once-through system.

Coal combustion as a source of heat has several advantages. For example, it results in **coal slag**, which under magneto hydrodynamic conditions is molten and provides a layer that covers all of the insulator and **electrode walls**. The electrical conductivity of this layer is sufficient to provide conduction between the gas and the electrode structure but not so high as to cause significant leakage of electric currents and consequent power loss. The reduction in thermal losses to the walls because of the slag layer more than compensates for any electrical losses arising from its presence. Also, the use of a seed material in conjunction with coal offers environmental benefits. In particular, the recombination chemistry that occurs in the duct of an MHD generator favours the formation of **potassium sulfate** in the combustion of **high-sulfur coals**, thereby reducing sulfur dioxide emissions to the atmosphere. The need to recover seed material also ensures that a high level of **particulate removal** is built into an MHD coal-fired plant. Finally, by careful design of the boiler and the combustion controls, low levels of nitrogen oxide emissions can be achieved.

2.2　Other MHD systems

In addition to coal as a fuel source, more-exotic MHD power generation systems have been proposed. Conventional nuclear reactors can employ hydrogen, or a noble gas such as argon or helium, as the working fluid, but they operate at temperatures that are too low to produce the thermal ionization used in MHD generators. Thus, some form of **nonequilibrium ionization** using seeding material is necessary.

In theory, solar concentrators can provide thermal energy at a temperature high enough to provide thermal ionization. Thus, solar-based MHD systems have potential, provided that **solar collectors** can be developed that operate reliably for extended periods at high temperatures.

The need to provide large pulses of electrical power at remote sites has stimulated the development of pulsed MHD generators. For this application, the MHD system basically consists of a rocket motor, duct, magnet, and connections to an electrical load. Such generators have been operated as sources for pulse-power electromagnetic sounding apparatuses used in geophysical research. Power levels up to 100 megawatts for a few seconds have been achieved.

A variation of the usual MHD generator employs a liquid metal as its **electrically conducting medium.** Liquid metal is an attractive option because of its high electrical conductivity, but it cannot serve directly as a thermodynamic working fluid. The liquid has to be combined with a driving gas or vapour to create a two-phase flow in the generator duct, or it has to be accelerated by a **thermodynamic pump** (often described as an ejector) and then separated from the driving gas or vapour before it passes through the duct. While such liquid metal MHD systems offer attractive features from the viewpoint of electrical machine operation, they are limited in temperature by the properties of liquid metals to about 1,250 K (about 975 ℃, or 1,800 ℉). Thus, they compete with various existing energy-conversion systems capable of operating in the same temperature range.

The use of MHD generators to provide power for spacecraft for both **burst and continuous operations** has also been considered. While both chemical and nuclear heat sources have been investigated, the latter has been the preferred choice for applications such as supplying electric **propulsion power** for deep-space probes.

Source：

1. http://www.iccf11.org/nuclear-accidents.

2. http://www.iccf11.org/radioactive-waste.

New words and phrases

abundance   *n.*   丰富；充裕

char   *n.*   炭

combustor   *n.*   （喷射引擎的）燃烧室

conductivity   *n.*   传导性；传导率；电导率

coolant   *n.*   冷冻剂；冷却液；散热剂

graphite   *n.*   石墨；黑铅

incompressible   *adj.*   不能压缩的

intercooler   *n.*   中间冷却器

insulator   *n.*   绝缘体

ratio   *n.*   比；比率

turbogenerator   *n.*   涡轮式发电机

atomized subcooled liquid droplet   雾化过冷液滴

burst and continuous operation   突发行动和持续行动

coal slag   煤渣

electrical conductivity   导电率;导电性

electrically conducting medium   导电介质

electrode wall   电极壁

exhaust gase   废气;排放气体

external load voltage   外部负载电压

high-sulfur coal   高硫煤

magnetic field strength   磁场强度

nonequilibrium ionization   非平衡电离

particulate removal   微粒清除

the positive and negative ions   正负离子

potassium carbonate   碳酸钾

potassium sulfate   硫酸钾

propulsion power   推进功率;驱动功率

solar collector   太阳能集热器

steam turbine-alternator unit   蒸汽透平发电机组

take shape   成形;具体化

the duct of boron nitride   氮化硼管

thermal ionization   热电离

thermodynamic pump   热力泵

Notes

1*   10 S/m   10 西/米。

Exercises

Please discuss over the following questions with your teammates after reading the text. You can support your view with more information from online or other channels.

1. Tell your teammates how many types of MHD systems according to the structure

and the kinds of classification.

2. What are the characteristics of closed-cycle MHD systems?

3. What are the advantages and limitations of liquid metal system?

Keys to Exercises

4. Can you give out the advantages of the coal-fired MHD systems?

Unit 9
Bioenergy

Text A   Bioenergy

Due to **volatile** and rising energy prices as well as increasing world wide energy demand, **bioenergy** is seen by many nations as an attractive **alterative** or addition to meet their current and future energy needs.

Many countries acknowledge bioenergy as a way to diversify their current energy mix, reduce dependency on fossil fuels such as oil and reduce greenhouse gas

emissions.

Currently, **biomass** residues and wastes already provide about 14 percent of the world's primary energy supplies, with the poential to meet up to half of world energy demands during the next century.

However, bioenergy does not come without its own pitfalls. Without proper research, production, management, bioenergy can increase pressure on already strained nature resources leading to economic loss, and reduced quality of life.

Bioenergy is also pointed out as a contributor to rising food prices and a threat to food security. There are good **biofuels** and bad biofuels, and, with good land and water management, there are good opportunities for many developing countries to produce their own transport fuels as well as food and fibre.

1. What Is Bioenergy?

Bioenergy is a renewable energy made available from materials derived from biological sources. Biomass is any organic material which has stored sunlight in the form of chemical energy. As a fuel, it may include wood, wood waste, straw, manure, sugarcane, and many other by-products from a variety of agricultural processes.

In its most narrow sense, it is a synonym to biofuel, which is the kind of fuel derived from biological sources. In its broader sense, it includes biomass, the biological material used as a biofuel, as well as the social, economic, scientific and technical fields associated with using biological sources for energy. This is a common misconception, as bioenergy is the energy extracted from the biomass, as the biomass is the fuel and the bioenergy is the energy contained in the fuel. Bioenergy is considered as the most **flexible** and extensive renewable source of energy integrated with other business manufacturing co-products.

Many energy and bioproducts research institutes have already done much work on renewable source of energy. For example, **the Energy and Bioproducts Research Institute at Aston University** conducts world-class research into all aspects of bioenergy including energy from waste and the development of new bioproducts and services, all of which contribute to improving the UK's carbon footprint and improving security of energy supplies.

Innovative research and development of bioenergy technologies has taken place at

Aston University for over 40 years. EBRI acts as a focus for international activities on scientific and technological aspects of biomass conversion and **utilisation** of products for renewable power, heat, transport fuels, hydrogen and chemicals. The institute works to extract the maximum value from all types of biomass, including waste resources.

EBRI is led by Professor Patricia Thornley, who also leads the **UK-wide Supergene Bioenergy Hub project**.

What is biomass?

Biomass is organic matter derived from living, or recently living organisms. Biomass can be used as a source of energy and it most often refers to plants or plant-based materials which are not used for food or feed, and are specifically called **lignocellulosic biomass**. As an energy source, biomass can either be used directly via combustion to produce heat, or indirectly after converting it to various forms of biofuel. Conversion of biomass to biofuel can be achieved by different methods. By 2010, there was 35 GW of globally installed bioenergy capacity for electricity generation, of which 7 GW was in the US.

There are many kinds of biomass. Many plants are used to produce the biomass, such as corn, willow, and sugarcane and hemp palm. Dead animals, waste oils and rubbish can also produce the energy.

(1) Agriculture waste

Many farmers already produce biomass energy by growing corn to make ethanol. But biomass energy comes in many forms. Virtually, all plants and **organic** wastes can be used to produce heat, power, or fuel.

Agricultural activities generate large amounts of biomass residues. While most crop residues are left in the field to reduce erosion and recycle nutrients back into the soil, some could be used to produce energy without harming the soil. Other wastes such as **whey** from cheese production and manure from livestock operations can also be profitably used to produce energy while reducing disposal costs and pollution.

Biomass energy has the potential to supply a significant portion of America's energy needs, while **revitalizing rural economies**, increasing energy independence, and reducing pollution. Farmers would gain a valuable new outlet for their products. Rural communities could become entirely self-sufficient when it comes to energy, using locally grown crops and residues to fuel cars and tractors and to heat and power homes

and buildings.

Opportunities for biomass energy are growing. For example, in the US, several million dollars of federal incentives are available through the 2002 Farm Bill to develop advanced technologies and crops to produce energy, chemicals, and other products from biomass. A number of states also provide incentives for biomass energy.

(2) Forest resources

Wood processing yields byproducts and waste streams that are collectively called wood processing residues and have significant energy potential. For example, the processing of wood for products or pulp produces unused sawdust, bark, branches, and leaves or needles. These residues can then be converted into biofuels or bioproducts. Because these residues are already collected at the point of processing, they can be convenient and relatively inexpensive sources of biomass for energy.

(3) **Livestock**

Algae as **feedstocks** for bioenergy refers to a diverse group of highly productive organisms that include microalgae, macroalgae (seaweed), and cyanobacteria (formerly called "blue-green algae"). Many use sunlight and nutrients to create biomass, which contains key components — including **lipids**, proteins, and carbohydrates — that can be converted and upgraded to a variety of biofuels and products. Depending on the strain, algae can grow by using fresh, saline, or **brackish water** from surface water sources, groundwater, or seawater. Additionally, they can grow in water from second-use sources, such as treated industrial wastewater; municipal, agricultural, or aquaculture wastewater; or produced water generated from oil and gas drilling operations.

(4) Municipal solid waste

Municipal solid waste (**MSW**) resources include mixed commercial and residential garbage, such as **yard trimmings**, paper and paperboard, plastics, rubber, leather, textiles, and food wastes. MSW for bioenergy also represents an opportunity to reduce residential and commercial waste by diverting significant volumes from landfills to the refinery.

(5) **Sanitary sewage and industrial organic wastewater**

Wet waste feedstocks include commercial, institutional, and residential food wastes (particularly those currently disposed of in landfills); organic-rich **biosolids**

(i. e. , treated sewage sludge from municipal wastewater); **manure slurries** from concentrated livestock operations; organic wastes from industrial operations; and biogas (the gaseous product of the decomposition of organic matter in the absence of oxygen) derived from any of the above feedstock streams. Transforming these "waste streams" into energy can help create additional revenue for rural economies and solve **waste-disposal problems**.

2. How Is Bioenergy Generated?

The amount of bioenergy depends on biomass. Conversion of biomass to biofuel can be achieved by different methods which are broadly classified into thermal, chemical and biochemical methods.

There are some examples of the usage.

First, simple use of biomass fuel — solid biomass. Combustion of wood for heat. The most conventional way in which biomass is used still relies on direct incineration. The most simple way to get biomass, is to burn woods.

Second, sewage biomass. A new bioenergy sewage treatment process aimed at developing countries is now on the horizon; the Omni Processor is a self-sustaining process which uses the sewerage solids as fuel to convert sewage waste water into drinking water and electrical energy.

Nowadays we develop new technologies to produce more energy using biomass:

Direct-combustion technology;

Biotransformation technology;

Thermochemical conversion;

Liquefaction technology;

Energy treatment technology for organic waste.

3. The Application of Bioenergy

Our use of fossil fuels is unsustainable. See Figure 9-1 and think why bioenergy is the answer to our fossil fuels crisis and reducing environmental damage, and why we need to hurry to make change.

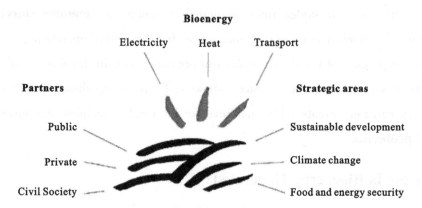

Fig. 9-1 The application of bioenergy

Firstly, bioenergy is not widely used in the world nowadays.

Through implementating the positive tax policy and application, the "Green Electricity Certificates System", the bioenergy production and application technologies came to maturity gradually in Sweden: integrated forest biorefinery[1]* produced lots of bioenergy in the form of heat, electricity and fuel particles; many heat and power plants used energy plant Salix as biomass fuels... In reality though, with the exception of the US, Brazil and some European countries, production of modern bioenergy and more specifically liquid biofuels around the world is still limited, especially in the case of Africa where the sector is still in its infancy.

Secondly, bioenergy is used to meet international energy crisis.

It is an important way to develop bioenergy for the international energy crisis. But such motivations are not the real points in concerning countries. As to motivations, it is more like a competitive method to grab renewable energy market in such developed countries as America and European Union. But in a few developing countries such as Brazil, it is just a method to evade trade protection of farm products and change competitive form and field. As to protection and support measures concerning with the biomass energy industry, it is a special form for such countries to carry out protection on farm products trade and to support agriculture. Distorted motives and measures of support and protection put great pressure on food security in the world, and it will worsen the development environment of most developing countries' economy.

At last, many countries depend on bioenergy more and more.

Bioenergy has the potential to add revenue of $ 6 billion per year to the New Zealand's economy. Bioenergy makes use of the residues and co-products of other business and manufacturing processes. It adds value by using residues which otherwise would be wasted. Increasing our use of bioenergy means a cleaner environment, a stronger economy and more employment. Investing in bioenergy is an investment in country's future.

Global Bioenergy Partnership (GBEP)[2*] was founded in 2006 on the idea that bioenergy can significantly contribute to energy access and security, climate change mitigation, food security, and ultimately sustainable development. The last decade has marked considerable growth for the organization, now counting with more than 70 members and an expanded number of activities in different countries.

GBEP brings together public, private and civil society stakeholders in a joint commitment to promote bioenergy for sustainable development. The Partnership focuses its activities in three strategic areas: sustainable development, climate change, and food and energy security.

4. Distribution of Bioenergy in the World

Biomass energy resources are rich and widely distributed. Global biomass potential energy use up to 350 EJ/year, according to WWF.

However, biomass is not balanced in the world, neither is the bioenergy. There are regional difference of electricity generation from biomass around the world.

4.1 Electricity from sugarcane bagasse in Brazil

The production process of sugar and ethanol in Brazil takes full advantage of the energy stored in sugarcane (see Fig. 9-2). Part of the **bagasse** is currently burned at the mill to provide heat for **distillation** and electricity to run the machinery.

Presently, it is economically viable to extract about 288 MJ of electricity from the residues of one tonne of sugarcane, of which about 180 MJ are used in the plant itself. Thus a medium-size distillery processing 1,000,000 tonnes (980,000 long tons; 1,100,000 short tons) of sugarcane per year could sell about 5 MW (6,700 Hp) of **surplus** electricity. At current prices, it would earn US $ 18 million from sugar and ethanol sales, and about US $ 1 million from surplus electricity sales. With advanced boiler and turbine technology, the electricity yield could be increased to 648 MJ per tonne of sugarcane, but current electricity prices do not justify the necessary investment. (According to one report, the World Bank would only finance investments

Fig. 9-2　The sugarcane in Brazil

Sugarcane (saccharum officinarum) plantation ready for harvest, Ituverava, São Paulo State, Brazil

in bagasse power generation if the price were at least USD 19/GJ or USD 0.068/kWh.)

4.2　Bioenergy in China

According to China's "Renewable Energy Medium and Long Term Development Plan" statistics, China's biomass energy is obviously much less invested in technology. For the development of biomass energy, first of all, we should unify the thinking from the upper level and raise understanding of the importance of biomass energy. We should also get to know the uneven **distribution** of bioenergy in China.

China is the world's third-largest producer of fuel ethanol, after the United States and Brazil (from RFA). China has set the goal of attaining one percent of its renewable energy generation through bioenergy in 2020. The development of bioenergy in China is needed to meet the rising energy demand.

Several institutions are involved in this development, of which the most notably are the Asian Development Bank and China's Ministry of Agriculture. There is also an added incentive to develop the bioenergy sector which is to increase the development of the rural agricultural **sector**. As of 2005, bioenergy use had reached more than 20 million households in the rural areas, with methane gas as the main biofuel. More than 4,000 bioenergy facilities produce 8 billion cubic meters every year of methane gas. By 2006, 20% of "gasoline" consumed was actually a 10% **ethanol-gasoline blend**. The annual use of **methane gas** was 19 cubic kilometers in 2010 (from *People's Daily*

Online). Electricity generation by bioenergy is expected to reach 30 GW by 2020.

Although only 0. 71% of the country's **grain yield** (3. 366 million tons of grain) in 2006 was used for production of ethanol, concern had been expressed over potential conflicts between demands for food and fuel, as crop prices rose in late 2006.

4.3　The forests in Amazon

The amount and spatial distribution of forest biomass in the Amazon basin (see Fig. 9-3) is a major source of uncertainty in estimating the flux of carbon released from land-cover and land-use change. Direct measurements of aboveground live biomass (AGLB) are limited to small areas of forest inventory plots and site-specific **allometric equations** that cannot be readily generalized for the entire basin. Furthermore, there is no **spaceborne** remote sensing instrument that can measure tropical forest biomass directly.

Fig. 9-3　The forests in Amazon

To determine the spatial distribution of forest biomass of the Amazon basin, a method based on remote sensing metrics representing various forest structural **parameters** and environmental variables was reported, with more than 500 plot measurements of forest biomass distributed over the basin. A decision tree approach was used to develop the spatial distribution of AGLB for seven distinct biomass classes of lowland old-growth forests with more than 80% accuracy. AGLB for other vegetation types, such as the woody and **herbaceous savanna** and secondary forests, was directly estimated with a regression based on satellite data.

Results show that AGLB is highest in Central Amazonia and in regions to the east and north, including the **Guyana**[3*]. Biomass is generally above 300 Mg ha^{-1} here except in areas of intense logging or open floodplains. In Western Amazonia, from the lowlands of Peru, Ecuador[4*], and Colombia to the Andean mountains[5*], biomass ranges from 150 to 300 Mg ha^{-1}. Most transitional and seasonal forests at the southern and northwestern edges of the basin have biomass ranging from 100 to 200 Mg ha^{-1}. The AGLB distribution has a significant correlation with the length of the dry season. We estimate that the total carbon in forest biomass of the Amazon basin, including the dead and belowground biomass, is 86 Pg C with ±20% uncertainty. Therefore, the bioenergy from the forest is mainly here.

5. The Advantages and Disadvantages of Bioenergy

Although, each source of biomass represents a technological challenge, for example, the diversity of raw materials will allow the decentralization of fuel production with geopolitical, economical and social benefits. The following passage presents a global overview of bioenergy, highlighting different feedstocks already in use, their qualities and limitations, also suggesting other potential ones.

There was a desire from our business community to do more. What are the advantages and disadvantages of bioenergy? If you take a closer look, you'll discover that it could be all right.

5.1 The advantages of bioenergy

(1) It's a renewable energy derived from agriculture.

The carbon dioxide (CO_2) released from biomass during production of bioenergy is from carbon that circulates the atmosphere in a loop through the process of **photosynthesis** and decomposition. Therefore, production of bioenergy does not contribute extra CO_2 to the atmosphere.

The amount of bioenergy depends on biomass. Biomass is organic matter derived from living, or recently living organisms. Biomass can be used as a source of energy and it most often refers to plants or plant-based materials which are not used for food or feed, and are specifically called lignocellulosic biomass. As an energy source, the biomass is from agriculture. Bioenergy is a renewable energy that can generate many additional benefits, which makes it a good alternative compares to other fuels like fossil fuels.

(2) By using bioenergy, we can reduce greenhouse gas emissions.

The extent of greenhouse gas (GHG) emissions reduction varies widely and depends on many factors including the biomass (feedstocks) used, how they are produced and the type and efficiency of the technology used to produce bioenergy. Generally, the reduction of GHG emissions from bioenergy systems is greatest where waste biomass is converted to heat or combined heat and power in modern plants located near to where the waste is generated.

Benefits from GHG reduction are potentially greater in bioenergy than those of other renewables. For example, **stubble** that is destined to be burnt in the field, can be harvested and combusted in an emission-controlled bioenergy plant. Hence, GHG emission reductions are made twice — once in the field through reduced burning and again through fossil fuel substitution from bioenergy production.

Considerable research is underway around the world to quantify the total lifecycle impacts of various bioenergy and other renewable energy systems. For example, through the IEA Bioenergy Task 38 Project — Greenhouse Gas Balances of Biomass and Bioenergy Systems.

(3) It can generate heat and electricity simultaneously.

Unlike most other renewable energy sources, biomass can generate both heat and electricity in a combined heat power (CHP) plant. This can then be used for a range of heating and cooling applications in industry, or for small communities.

5.2 The disadvantages of bioenergy

(1) Inefficient as compared to fossil fuels.

It doesn't get close to fossil fuels in regards to efficiency. In fact, some renewable sources of energy like biofuels are **fortified** with fossil fuels to increase their efficiency.

(2) Risk of deforestation.

If deforestation is allowed to happen, scores of animal and bird species would be rendered homeless, not to mention the drought as a result. In fact, this is the main reason for slowing down the large scale use of biomass fuel. Governments feel replanting efforts may not match the rate of cutting down of trees.

(3) Not entirely clean.

Using animal and human waste to power engines may increase methane gases, which are harmful to the Earth's ozone layer.

(4) Water shortage.

Expansion of arable land areas will require more irrigation. Water depletion might occur in the countries producing feedstocks for biofuels. A new bioenergy sewage treatment process aimed at developing countries is now on the horizon; the Omni Processor is a self-sustaining process which uses the sewerage solids as fuel to convert sewage waste water into drinking water and electrical energy.

6. Future Development of Bioenergy

Watch Figure 9-4 to predicate the future development of bioenergy.

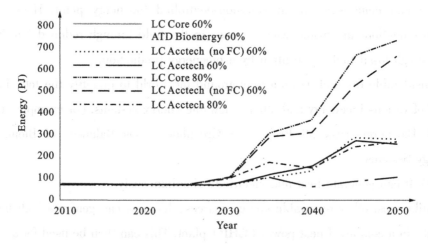

Fig. 9-4 The future development of bioenergy

The future development of bioenergy will be rapid and wide. All the countries are taking it more seriously.

In China, the development and use of new energy has attracted extensive attention. China's experts conducted important research. The important functions of biomass to solve energy crisis, issue concerning agriculture, countryside and farmers, ecology destroy, environment pollution and sustainable development were analyzed in the paper. China biomass development way and goal were planed and suggested, the status of bioenergy and biomass utilization at home and abroad were also summarized.

Among renewable energy resources, bioenergy is one of the fastest growth energy alternatives with tremendous potential in China. The thermal, physical, and biological processes of conversion of biomass yield a number of products and can be obtained as gases, liquids, solid fuels, and electricity as well as a variety of chemicals. Various bioenergy technologies that have been developed are at the fundamental research, demonstration, and commercialization stages.

The experts explain the important roles bioenergy plays in China. Firstly, the application status of bioenergy technologies are introduced, including biogas, fuel ethanol, biodiesel, and power generation at the commercialization stage. Secondly, the current research progresses are analyzed of ethanol derived from lignocellulose, sweet sorghum and cassava, biodiesel from jatropha, biomass **briquetting**, synthesized fuels and pyrolysis technologies at the fundamental research and demonstration stages. Finally, it is concluded that the key areas for developing bioenergy for the future are the exploitation of new biomass resources and R&D in biofuels from non-food biomass resources, as well as the development of commercialization methods suitable for developing countries.

Large amounts of small-scale bioenergy projects were carried out in China's rural areas in light of its national renewable energy policies. These projects applied **pyrolysis gasification** as the main technology, which turned biomass waste at low costs into biogas. We can select seven bioenergy projects in Shandong Province as a case and assesses these projects in terms of economy, technological performance and effectiveness. Results show that these projects have not achieved a satisfying performance after 10 years experience. Many projects have been discontinued. This failure is attributed to a complex of shortcomings in institutional structure, technical level, financial support and social factors. For a more successful future of bioenergy development in rural areas, China should reform its institutional structure, establish a renewable energy market and enhance the technological level of bioenergy projects.

In New Zealand, bioenergy is the best energy of the future. It is the most used form of renewable energy globally — wider than hydroelectricity or wind, providing heat, electricity and fuel for transport and has the potential to create a major industry for New Zealand. Increasing use of bioenergy is good for the economy — particularly in rural New Zealand — providing economic growth through improved land use, new businesses and employment that will strengthen rural communities.

The Bioenergy Association promotes and coordinates the expansion of the New Zealand bioenergy sector and helps its members' bioenergy businesses grow and flourish. More than 10 percent of New Zealand's energy currently comes from bioenergy.

In a word, bioenergy will be used widely in the future in the world.

Source:

1. https://www.wikipedia.net.

2. https://www.nature.com/articles/ncomms1316.

3. http://en.cnki.com.cn/Article_en/CJFDTotal-GLJH201306040.htm.

 New words and phrases

alternative　*n.*　可供选择的事物

　　　　　　adj.　可供替代的；非传统的；另类的

bagasse　*n.*　（当作燃料或造纸用的）甘蔗渣

bioenergy　*n.*　生物能（源）

biofuel　*n.*　生物燃料

biomass　*n.*　（单位面积或体积内的）生物量

biosolids　*n.*　生物固体（从处理过的废水中取得的固体或半固体物质，常用作肥料）

briquet　*v.*　把……压制成块

cogeneration　*n.*　废热发电

distribution　*n.*　分布

distillation　*n.*　蒸馏（过程）；精华

feedstock　*n.*　原料

flexible　*adj.*　能适应新情况的；灵活的；可变动的；柔韧的；可弯曲的；有弹性的

fortify　*v.*　加强；（加入某物）强化

innovative　*adj.*　革新的；创新的

lipid　*n.*　（生化）脂质；脂肪

livestock　*n.*　牲畜；家畜

organic　*adj.*　有机（体）的

parameter　*n.*　规范；范围；因素；特征

photosynthesis　*n.*　光合作用

sector　*n.*　区

spaceborne　*adj.*　空运的;飞船上的;在宇宙间的

stubble　*n.*　作物收割后遗留在地里的残茎,茬

surplus　*adj.*　过剩的

utilisation　*n.*　利用

volatile　*adj.*　易变的,不稳定的

whey　*n.*　乳清

allometric equation　相对关系式;开度量方程

biotransformation technology　生物转化技术

brackish water　半咸水;微咸水

direct-combustion technology　直接燃烧技术

energy treatment technology for organic waste　有机垃圾能源化处理技术

ethanol-gasoline blend　乙醇-汽油混合物

grain yield　粮食产量

herbaceous savanna　草本稀树草原

lignocellulosic biomass　木质纤维素生物质

liquefaction technology　液化技术

manure slurry　粪肥浆

methane gas　沼气

municipal solid waste（MSW）　城市固体废物

pyrolysis gasification　热解气化

revitalize rural economies　振兴农村经济

sanitary sewage and industrial organic wastewater　生活污水和工业有机废水

the Energy and Bioproducts Research Institute at Aston University　阿斯顿大学能源和生物产品研究所

the UK-wide Supergen Bioenergy Hub project　全英超原生物能源枢纽项目

thermochemical conversion　热化学转化技术

waste-disposal problem　废物处理问题

yard trimming　庭院装饰

Notes

1* integrated forest biorefinery（IFBR） 联合型林产品生物精炼厂,或称森林生物质能综合提炼厂。

2* Global Bioenergy Partnership（GBEP） 全球生物能源伙伴关系。成立于2006年,其理念是生物能源可大大促进能源获取及其安全、减缓气候变化、促进粮食安全,最终实现可持续发展。

3* Guyana 圭亚那,拉丁美洲国家。

4* Ecuador 厄瓜多尔,南美洲西北海岸的国家。

5* Andean mountains 安第斯山,位于南美洲。

Exercises

I. **Read the text and discuss over the following questions with your partner.**

1. What is bioenergy? Please describe it according to the passage.

2. What do you think is the future of bioenergy in China?

3. How to avoid the negative impacts of bioenergy?

II. **Please give the Chinese or English equivalents of the following terms.**

1. diversify their current energy

2. organic materia

3. biological source

4. all aspects of bioenergy

5. innovative research and development of bioenergy technologies

6. revitalize rural economies

7. brackish water

8. municipal solid waste

9. biotransformation technology

10. liquefaction technology

11. 逃避农产品贸易保护

12. 燃烧生物热电联产项目

13. 为蒸馏提供热量

14. 毁林风险

15. 将受污染废水转化为饮用水

16. 生物能源的未来发展

17. 用化石燃料加强

18. 产生热和电

19. 可再生能源

20. 广义分类

III. Please translate the following sentences into English.

1. 生物能源是从生物质中获得的一种可再生能源。

2. 从最狭义的意义上说,它是生物燃料的同义词,生物燃料是从生物资源获得的燃料。

3. 生物质可以作为一种能源,它通常指不用作食物或饲料的植物或植物基材料,也称为木质纤维素生物质。

4. 生物质是任何以化学能形式储存阳光的有机物质。

5. 以发展中国家为对象的新的生物能源污水处理程序现在即将启动。

6. 由于能源价格的波动和上涨以及全世界能源需求的增加,生物能源被许多国家视为一种有吸引力的替代物或补充物,以满足它们目前和未来的需求。

7. 许多国家承认使用生物能源是使其目前的能源组合多样化、减少对石油等化石燃料的依赖和减少温室气体排放的一种方式。

8. 巴西的糖和乙醇生产中充分利用了甘蔗中储存的能量。

9. 中国制定了 2020 年通过生物能源实现可再生能源发电的 1% 的目标。

10. 因此,生物能源的生产不会像化石燃料那样对大气产生额外的二氧化碳。

IV. Please translate the following passage into Chinese.

This paper presents a strategy for China's rural energy development at the turn of this century. It proposes the establishment of a three-dimensional and open system of rural energy engineering proceeding from an analysis of China's rural energy resources, the state of rural development, the supply of energy in the countryside and other exogenous variables. The system should consist of lateral and vertical networks of energy supply encompassing all rural areas, its mainstay being small hydroelectric power stations, small coal mines and biogas pits on farms.

By using the word "mainstay", the author does not intend to describe these as being able to satisfy all energy demands, instead, means to regard them as the major sources of energy for China's countryside since they are reliable and easy to set up and require short development cycles while great in sheer numbers.

The networks of energy supply cover solar energy and wind power as well as commodity energy supplied by the State. Nevertheless, it will be difficult to significantly raise the proportion of commodity energy to the total rural energy consumption by the end of this century due to the limited scale of its production and the inadequate technical resources for developing new energy sources. This "mainstay" described in this paper can, in an indirect manner, help improve the country's ecological balance and boost the development of ecological energy.

V. Please read the passage and get to know something about the biomass gasifiers. Then tell your partner what this new type of biomass gasifier is.

Biomass Gasifiers: from Waste to Energy Production

Biomass gasifiers utilize a proven principle to convert unused agricultural and forestry residues into a clean, high-quality gas. The principle was first used during World War II. This gas can replace conventional energy sources, such as fuel oil and natural gas for crop drying, space heating, and industrial boilers. And the biomass gasifiers can also be used directly to drive most internal combustion engines. The biomass gasifiers can operate on various fuels. And the fuels are usually considered unusable or low-value wastes, such as sawdust, wood chips, corn cobs, nut shells, rice hulls, etc. Gas produced by down-draft gasifiers is tar-free.

Text B The US Firm Develops Next Generation Biofuel

The development of second-generation biofuels — those that do not rely on grain crops as inputs — will require a diverse set of feedstocks that can be grown sustainably and processed cost-effectively. Here we review the outlook and challenges for meeting hoped-for production targets for such biofuels in the United States.

The importance of renewable biofuels in displacing fossil fuels within the transport sector in the United States is growing, especially in the light of concerns over energy security and global warming. The US federal government, as well as most governments worldwide, is strongly committed to displacing fossil fuels with renewable, potentially low carbon biofuels produced from biomass. The primary motivation for these efforts is both to decrease reliance on fossil fuels, particularly imported fuels, and to address concerns over the contribution of fossil-fuel consumption by the transport sector to global warming. The US federal government has therefore set a target of displacing 30% of current US gasoline (petrol and diesel) consumption within the transportation sector with biofuels by 2030. The European Union, China, Australia and New Zealand have also established similar targets for biofuel production.

Currently, the majority of biofuel production in the United States is in ethanol derived from starch or grain-based feed stocks, such as corn (maize). Sugarcane is also a prime resource for biofuel production in Brazil and other regions of the world. These biofuels will be produced through the conversion of lignocellulosic biomass and are commonly referred to as second-generation biofuels. Those biomass feedstocks are not primarily composed of starches, but rather of the complex matrix of **polysaccharides** and **lignin** that forms **plant cell walls**. These lignocellulosic materials are inherently more difficult than grain-based materials to convert into **fermentable sugars** (see Fig. 9-5). The plant cell walls found within lignocellulosic biomass are a complex mixture of polysaccharides, pectin and lignin. The polysaccharides are chemically linked to the lignin, and these complexes are very recalcitrant to processing and **depolymerization** into their respective monomers.

To meet these production targets, a robust and sustainable supply of the requisite feedstocks must be developed and established. A joint study by the US Departments of

Energy and Agriculture, often referred to as the "Billion Ton Study", determined that roughly 1. 18 billion tonnes of non-grain biomass feedstocks could be produced on a renewable basis in the United States each year and dedicated to biofuel production. These feedstocks are primarily distributed among forestry and agricultural resources. Assuming a conservative estimate of biofuel production at 190 liters (50 gallons) per dry tonne, this would create an upper limit of biofuel production, albeit a highly optimistic one to be achieved over this time period, of 247 billion liters (65 billion gallons) per year.

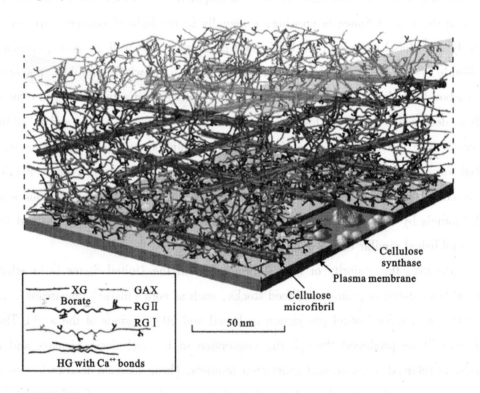

Fig. 9-5 The chemical and structural complexities of the plant cell wall

Source: https://genomeblogy. ciomedcentral. com/articles/10. 1186/gb-2008-9-12-242.

1. Forestry Resources

A recent report reported that the amount of forestland, as of 2002, in the United States was roughly 303 million hectares (750 million acres). This represents one-third of the total land area of the nation. The majority of these lands are held by the forestry industry or other private interests. **A significant portion** of this land is not accessible to forestry equipment, however. In addition, approximately 68 million hectares of forestland is not considered as a viable biofuel feedstock growth area. Current forest product manufacturing techniques produce large amounts of mill residues, known as secondary residues. These secondary residues account for approximately 50% of current biomass energy consumption in the United States, and will continue to play a vital role in producing biofuels. In total, the amount of harvested and consumed forestry resources in the United States — 127. 8 million dry tonnes — is considerably less than the available inventory. This excess capacity indicates that there is a significant amount of forestry resources — 331 million dry tonnes — that could be dedicated to biofuel production on a sustainable basis (see Fig. 9-6).

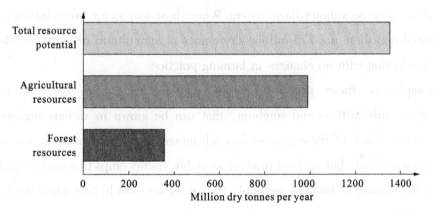

Fig. 9-6   Estimates of biomass available for conversion into biofuels per year within the United States

Some of the leading candidates that could be grown on these lands specifically for biofuel production are **hybrid poplar, eucalyptus, loblolly** pine, willow and silver maple. Poplar has several characteristics that make it an attractive candidate biofuel feedstock: it can be grown in several temperate climates as a **short-rotation woody crop**; it grows relatively rapidly at high density; it is a good plantation tree; and it has a fully sequenced genome. Willow and loblolly pine are also strong short-rotation woody crop candidates. Eucalyptus, native to Australia but grown throughout the world, has been grown and studied extensively in California and Florida, and appears to be amenable to high-density cultivation in plantation farms.

Another key aspect to forestry-resource management is the biomass turn over from leaf litter. This phenomenon is an annual process for deciduous trees, and occurs after leaf senescence, when most of the reserves have been remobilized except for cell-wall polysaccharides.

2. Agricultural Resources

Agriculture is the third largest use of land in the United States, estimated at 182-184 million hectares. It was recently reported that approximately 141 million hectares of land are actively farmed to grow crops, with an additional 16 million hectares of idle cropland. These idle croplands include those that have been placed in **the Conservation Reserve Program** (CRP). Other uses include 27 million hectares for pasture. A significant area of cropland, 25 million hectares, uses **no-till cultivation** to reduce soil erosion and maintain soil nutrients, whereas another 20 million hectares of cropland use a conservation tillage system. When these factors are taken into account, it is estimated that there are 175 million dry tonnes of agricultural resources available for biofuel production with no changes in farming practice.

Examples of these **perennial** crops include herbaceous species, such as **switchgrass**, **miscanthus** and sorghum, that can be grown in various regions of the United States. Each of these grasses has advantages and disadvantages that must be carefully considered, but all hold promise as viable energy crops that could significantly increase the amount of biomass available for conversion into biofuel when implemented appropriately.

The inclusion of these perennial crops within agricultural resource lands or CRP land is projected to result in 14 or 22 million hectares associated with moderate (11 dry tonnes per hectare) and high (18 dry tonnes per hectare) yields, respectively. With a high percentage of these perennial crops dedicated to biofuel production, this scenario projects that 523 to 898 million dry tonnes of biomass could be produced at moderate and high yields, respectively. Crop residues remain the most significant component (50%) of the available biomass, with perennial crops contributing 30%-40%.

3. Genetics and Feedstock Improvement

In addition to growing currently available feedstocks on available land to produce biofuels, the realization of dedicated energy crops with enhanced characteristics would represent a significant step forward. The genetic sequences of a few key biomass feedstocks are already known, such as poplar, and there are more in the sequencing pipeline. This genetic information gives scientists the knowledge required to develop strategies for engineering plants with far superior characteristics, such as diminished **recalcitrance** to conversion.

There have been several recent examples where genetic engineering has been used to modify the composition of the plant in order to hypothetically reduce the cost associated with the conversion process. The presence of lignin in plant cell walls impedes the hydrolysis of polysaccharides to simple sugars. Lignin and its by-products can also inhibit the **microbes** that carry out fermentation, decreasing biofuel yield. Other examples include modifying lignin **biosynthesis** in plants in order to make the plant more readily broken down in the biorefinery, adjusting the types of lignin present in plants, and adjusting the ratio between polysaccharides and lignin. In addition to modifying the intrinsic polysaccharide / lignin composition and central **metabolism** of the feedstock itself, other research groups are attempting to express enzymes directly within plants that are capable of breaking down cellulose into glucose. These enzymes are called **cellulases**, and supplying them to the production process represents one of the largest costs in biofuel production. Expressing and localizing cellulases within the plant could potentially eliminate the need for producing the cellulase offline at the biorefinery. Researchers have successfully expressed the gene encoding the **catalytic domain** of one cellulase into arabidopsis, tobacco and potato.

4. Challenges for the Future

Numerous challenges must be **addressed** for feedstock production to reach established targets. Some of the main challenges are associated with developing a vast amount of acreage within the United States dedicated to feedstock growth for biofuel conversion, including ensuring sustainability, reducing cost, and devising responsible land-use change policies. In regard to agricultural residues, care must be taken to ensure that removal of the residues from the fields does not negatively impact any other **interlinked parameter**, such as silage and other established beneficial farming practices. The development of specialized harvesting equipment for these residues also needs to be addressed if gains in production are to be realized.

As dedicated non-food energy crops, most probably in the form of grasses and short-rotation woody crops, become widespread and grown on **marginal lands** or CRP, land management practices and crop selection controls must be established in order to minimize any indirect carbon or nitrogen emissions from the soil as a result of changes in land use.

Other concerns that must also be addressed are the development of the necessary infrastructure for harvesting, collecting, processing, and distributing large volumes of biofuels. This strategy will therefore require a means to distribute the biofuels from the points of production in the Midwest to the primary points of consumption in the populous West and East coasts.

In conclusion, the role of sustainable, cost-effective, and **scalable** feedstock production is one of the most pressing needs in the realization of a biofuels industry capable of replacing a significant portion of the fossil-fuel consumption of the United States. It is important to recognize that different feedstocks will need to be grown in different regions to meet the tonnage required. This diversification in the supply chain should be considered a strength and not a weakness, as the numerous possible feedstock and environmental combinations should be able to maximize productivity and sustainability while minimizing cost. Although enough hypothetical biomass seems to be available to meet biofuel production targets, **significant hurdles** remain before those numbers can become a cost-effective and environmentally beneficial reality. Genetic engineering and **synthetic biology** can be used to produce feedstocks with the desired traits, especially when leveraged with existing expertise within the plant biology and **agronomy** communities.

Source:

1. http://www.eia.doe.gov/emeu/steo/pub/special/2008_sp_02.html.

2. http://apps1.eere.energy.gov/news/news_detail.cfm/news_id=11633.

3. http://feedstockreview.ornl.gov/pdf/billion_ton_vision.pdf.

4. http://www.iccf11.org/nuclear-accidents/

5. http://www.iccf11.org/radioactive-waste.

New words and phrases

address   *v.*   设法解决;提出;演讲

agronomy   *n.*   农艺学;农学

biosynthesis   *n.*   生物合成

cellulase   *n.*   木纤维质酵素;纤维素酶

depolymerization   *n.*   解聚(合)作用

eucalyptus   *n.*   桉树

loblolly   *n.*   火炬松

lignin   *n.*   木质素

metabolism   *n.*   新陈代谢

microbe   *n.*   微生物;细菌

miscanthus   *n.*   芒属植物

perennial   *adj.*   终年的,长久的;(植物)多年生的

polysaccharide   *n.*   多糖

recalcitrance   *n.*   反抗;固执;抗性

scalable   *adj.*   可攀登的;可升级的;可扩展的

switchgrass   *n.*   柳枝稷

a significant portion   很大一部分

catalytic domain   催化区

fermentable sugar   可发酵糖

hybrid poplar   杂交杨树

interlinked parameter   相互关联的参数

marginal land   边际土地

no-till cultivation   免耕

plant cell wall   植物细胞壁

short-rotation woody crop   短伐期木本作物

significant hurdle   重大障碍

synthetic biology   合成生物学

the Conservation Reserve Program (CRP)   土地休耕计划

 Exercises

Please discuss over the following questions with your teammates after reading the text. You can support your view with more information from online or other channels.

1. Please generalize the factors of genetics and feedstock improvement.

2. What are the main resources to develop next-generation biofuel in the text?

3. Can you bring up the challenges for the future in the **Keys to Exercises** text? What do you think of them?

与本书配套的二维码资源使用说明

　　本书部分课程及与纸质教材配套数字资源以二维码链接的形式呈现。利用手机微信扫码成功后提示微信登录，授权后进入注册页面，填写注册信息。按照提示输入手机号码，点击获取手机验证码，稍等片刻收到4位数的验证码短信，在提示位置输入验证码成功，再设置密码，选择相应专业，点击"立即注册"，注册成功。（若手机已经注册，则在"注册"页面底部选择"已有账号？立即注册"，进入"账号绑定"页面，直接输入手机号和密码登录。）接着提示输入学习码，需刮开教材封面防伪涂层，输入13位学习码（正版图书拥有的一次性使用学习码），输入正确后提示绑定成功，即可查看二维码数字资源。手机第一次登录查看资源成功以后，再次使用二维码资源时，只需在微信端扫码即可登录进入查看。